Partial
Hospitalization
A Current Perspective

APPLIED CLINICAL PSYCHOLOGY
Series Editors:
Michel Hersen and Alan S. Bellack
University of Pittsburgh, Pittsburgh, Pennsylvania

PARTIAL HOSPITALIZATION: A Current Perspective
Edited by Raymond F. Luber

A Continuation Order Plan is available for this series. A continuation order will bring delivery of each new volume immediately upon publication. Volumes are billed only upon actual shipment. For further information please contact the publisher.

Partial Hospitalization
A Current Perspective

Edited by

Raymond F. Luber

Western Psychiatric Institute and Clinic
School of Medicine
University of Pittsburgh
Pittsburgh, Pennsylvania

Plenum Press · New York and London

Library of Congress Cataloging In Publication Data

Main entry under title.

Partial hospitalization.

Includes index.
1. Psychiatric hospital care. 2. Partial hospitalization. I. Luber, Raymond
F. [DNLM: 1. Day care. 2. Community mental health services. 3. Mental
disorders—Rehabilitation. WM29. 1 P273]
RC439.2.P37   616.8'91   78-31915
ISBN-13: 978-1-4613-2966-4   e-ISBN-13: 978-1-4613-2964-0
DOI: 10.1007/ 978-1-4613-2964-0

© 1979 Plenum Press, New York
Softcover reprint of the hardcover 1st edition 1979
A Division of Plenum Publishing Corporation
227 West 17th Street, New York, N.Y. 10011

To
Janice, Marty, and Jenna

Contributors

Marguerite Conrad, R. N., Ed.M., M.P.H., Late Assistant Professor of Nursing for Partial Hospitalization and Aftercare, Clinical Assistant Professor, Boston University School of Nursing, Boston, Massachusetts

Thomas Detre, M.D., Professor and Chairman, Department of Psychiatry, Western Psychiatric Institute and Clinic, University of Pittsburgh School of Medicine, Pittsburgh, Pennsylvania

Thad A. Eckman, Jr., Ph.D., Camarillo-UCLA, Neuropsychiatric Institute and California Lutheran College, Thousand Oaks, California

V. DeCarolis Feeg, R.N., M.A., Division of Biological Health, The Pennsylvania State University, University Park, Pennsylvania

F. Dee Goldberg, M.P.H., Deputy Commissioner, Program Support Service, Division of Mental Health, Columbus, Ohio

Michel Hersen, Ph.D., Professor of Clinical Psychiatry, Director, Resocialization Treatment Center, Western Psychiatric Institute and Clinic, University of Pittsburgh School of Medicine, Pittsburgh, Pennsylvania

David L. Kupfer, B.A., Department of Psychology, University of Georgia, Athens, Georgia

Benjamin B. Lahey, Ph.D., Assistant Professor, Department of Psychology, University of Georgia, Athens, Georgia

Paul M. Lefkovitz, Ph.D., Director, Partial Hospitalization, Gallahue Mental Health Center, Community Hospital of Indianapolis, Indianapolis, Indiana

Raymond F. Luber, M.Div., Assistant Professor of Clinical Psychiatry, Director, Partial Hospitalization, Department of Psychiatry, Western Psychiatric Institute and Clinic, University of Pittsburgh School of Medicine, Pittsburgh, Pennsylvania

John T. Neisworth, Ph.D., Associate Professor of Special Education, Department of Special Education, The Pennsylvania State University, University Park, Pennsylvania

Joan Perrault, B.S.N., M.P.H., Lecturer in Nursing, George Mason University, Consultant, Mental Health Program Development, Fairfax, Virginia

Samuel M. Turner, Ph.D., Assistant Professor of Psychology, Department of Psychiatry, Western Psychiatric Institute and Clinic, University of Pittsburgh School of Medicine, Pittsburgh, Pennsylvania

Stephen Washburn, M.D., Chief, Partial Hospitalization Service, McLean Hospital, Belmont, Massachusetts; Assistant Professor of Psychiatry, Harvard Medical School, Cambridge, Massachusetts

Foreword

There was a time, not long ago, when the only treatment options considered to be worthwhile for patients requiring psychiatric care were the 50-minute hour on the one hand, or full-time hospitalization on the other. Most of us were convinced in those days that treatment could, and indeed should, take place with a minimum of involvement by the patient's family. Nor did we really consider that the community in which a patient lived was a significant contributor to either his illness or its cure.

These naive assumptions were strongly challenged, of course, beginning with the questions of social psychiatrists in the 50s and continuing with the quiet growth of the patients' rights movement. Thus it is no mere coincidence that when the community psychiatry movement emerged in the mid-60s as a powerful force for profound change in our traditional practice, the concept of partial hospitalization, which can be traced back at least 30 years, became a symbol of the new social psychiatry. Partial hospitalization had singular advantages well attuned to the times: it did not force a separation between the patient and his family; it cost far less to deliver than inpatient care; and it avoided the stigma of institutionalization while still providing far more care than the traditional psychotherapeutic hour. In a few years' time, several well-controlled studies documented that virtually all patients who were customarily treated on an inpatient basis could be effectively managed and treated in a day hospital.

Like so many of their counterparts in other modes of health care delivery, the advocates of day hospitals and other partial hospitalization programs have often been somewhat reluctant to expose their innovative methods of patient care to the time-consuming process of close critical scrutiny, and as a result several important issues remain unresolved. Some of these issues bear upon the treatment itself, but others, just as important, involve the cost of such treatment to our society as a whole and to the members of the patients' families in particular.

While we readily assume that what is good for the patient and protects his or her rights is what should be done, we do not know what effect, if any, a severely depressed or schizophrenic patient living at home has on children or other relatives living in the same household, nor can we measure at this stage whether or not or under what circumstances the advantages of partial hospitalization are outweighed by the disadvantages caused by the patient's presence in the family.

It is also an historical truism that health care delivery systems rarely make an effort to address the question of efficacy until forced to do so hurriedly because time — or funding — is running out. As our increasingly cost-conscious society begins to demand proof, it becomes clear that amazingly little has been done to define specific uses of partial hospitalization or to identify what programs and applications within partial hospitalization are best suited to a specific group of patients. We still have no data to tell us whether a specific group of patients, such as ex-inpatients about to be discharged into the community, are best served by a partial program, or whether the improvement of patients' social skills or work habits attempted in partial hospitalization programs is maintained over time.

As a result, while the cost of treatment in these settings is at most half of what it would cost to admit the patient to a full-time inpatient unit, lacking guidelines and proof of efficacy, third party payers have never developed a consistent policy of reimbursement. Some, recognizing that patients can get care as effective as inpatient care for half the price, have shown willingness to underwrite the expenses, but there are many states where the reimbursement schedule is barely sufficient to provide a baby-sitting service for adults, which was very far from the aim of partial hospitalization programs.

This volume is among the first to take a long step forward toward bridging the gap between treatment planning and assessment data.

THOMAS DETRE, M.D.

Preface

This book is the result of several converging influences. One influence has been my own personal experience. In varying capacities over the past several years I have been associated with partial hospitalization. During this time, several changes (many of them unplanned and imposed by outside forces) have made their impact on partial hospitalization. Some of these changes have been provincial in nature while others have been the direct result of changes in national mental health care policies. Some changes were welcomed, others were met with violent opposition. Through it all, however, I have been impressed with and at times astonished by the resiliency and dedication of those affiliated with partial programs. It has been obvious that professionals in the field have considered the treatment modality to be utilitarian and worthwhile. It is gratifying that we are *beginning* to substantiate this "clinical impression" with empirical data.

A second influencing factor has been the obvious need for some attempt to integrate the available material related to partial hospitalization. The last 3 to 5 years have seen a marked increase in clinical and research publications related to partial hospitalization as well as a growth in professional organizations primarily concerned with this treatment modality. This has been an encouraging and long-needed development. Unfortunately, these efforts have had little unity and even less exposure. In this period of time I have received numerous requests from professionals, students, and other interested individuals for information about partial hospitalization; at times the requests were as simple as supplying a bibliography of resources. Unfortunately, even this minimum request was difficult to meet. It is hoped that this book will, therefore, meet two needs: (1) to provide a comprehensive survey and integration of the most significant clinical, conceptual, and research developments in the field of partial hospitalization; and (2) to provide an impetus and, ideally, some direction for further developments in these areas in the future.

A final influence has been the encouragement received from many professional associates and colleagues. This encouragement is gratefully acknowledged and much appreciated. It is my hope that the final product serves as an adequate response.

Many people deserve recognition for their contributions to this book: first, the authors of the various chapters, whose diligence, scholarship, and efforts will be immediately evident to the reader; second, Doris Miller, whose secretarial efforts made the work involved in completing the project less strenuous and burdensome than it might have been; third, a special friend and colleague, Michel Hersen, whose support and modeling behavior over the past few years have been of particular value; and finally, all those associated with and dedicated to the concept of partial hospitalization, without whom there would be little need for this book.

RAYMOND F. LUBER

Contents

PART IV: PROBLEMS AND FUTURE DIRECTIONS

Chapter 7
Treatment Orientation and Program Implications 139
Samuel M. Turner

Chapter 8
Patient Population and Treatment Programming 151
Paul M. Lefkovitz

Chapter 9
Funding Partial Hospitalization Programs 173
F. Dee Goldberg and Joan Perrault

Chapter 10
Future Directions of Partial Hospitalization 183
Raymond F. Luber

I

Treatment Approaches to Partial Hospitalization

Introduction

In the years following the passage of The Mental Retardation Facilities and Community Mental Health Centers Construction Act of 1963, which made partial hospitalization a mandated service in federally funded projects, the treatment modality has experienced its greatest period of growth. During this time there has been a 700% increase in the number of partial hospitalization programs and a 14-fold increase in the number of patients treated in these programs. Despite future uncertainties regarding funding sources, partial hospitalization is still experiencing expansion and growth in several areas.

However, despite this growth rate, partial hospitalization has been faced with a major problem: defining and implementing a consistent treatment program. One source of this particular problem has been the general lack of a unifying principle or rationale for programming. As a result, a confusing diversity in treatment approaches has evolved. These approaches range from highly structured to relatively unstructured program formats and have generally been based on theoretical or conceptual models supported by a minimum of empirical data.

In this section, the question of treatment orientation will be addressed. First, to place the problem in context, Raymond Luber describes the historical trends in the growth and expansion of partial hospitalization, presents some of the logical and conceptual advantages of the treatment modality, and defines some of the problems and important issues that have persisted in the area over the years. Thad Eckman then describes in detail a behavioral approach to treatment programming including relevant research data supporting the conceptual model. Finally, Stephen Washburn and Marguerite Conrad present a therapeutic community model for partial hospitalization programming.

The Growth and Scope of Partial Hospitalization

Raymond F. Luber

Introduction

With the initial wave of expansion in the partial hospitalization movement, which began about 1963, a concurrent mood of expectation and
optimism emanated from the proponents of this "new" treatment modality. This optimism, at times reaching near-grandiose proportions,
was reflected by Barnes (1964), who stated:

> The psychiatry of the nineteenth and early twentieth centuries was built
> around the large state hospital. After World War II emphasis began to pass to
> the "mental hygiene clinic" and the psychiatrist's office. The mid and late 60s
> will see the central position shifting to the day hospital and fewer and fewer
> cases of psychosis and severe neurosis will require residential hospitaliza-
> tion. . . . Thus, the "New Psychiatry" of the 60s will rest firmly on the day
> hospital. (p. x)

Despite substantial growth in both the number of partial hospitalization programs and the number of patients treated in such programs,
the day hospital has not become the primary mode of psychiatric intervention nor has it supplanted the inpatient service as the principal arena
of treatment. This undoubtedly reflects the fact that initial expectations
were based more on uncontrolled enthusiasm than on a realistic evaluation of the situation at the time. Consequently, over the past 15 years a

Raymond F. Luber • Department of Psychiatry, Western Psychiatric Institute and Clinic,
University of Pittsburgh School of Medicine, Pittsburgh, Pennsylvania 15261.

moderating trend has developed characterized by a gradually increasing interest in defining the distinctive role of partial hospitalization *within* the broad spectrum of psychiatric services.

Various facets of this effort to define and specify the role of partial hospitalization will be considered in the following chapters. This chapter will outline the historical development of partial hospitalization, describe its expansion (in total number of programs, number of patients treated, and treatment settings), describe some of the theoretical advantages of the modality, and, finally, summarize the problems that have persisted in the treatment form over the past 15 years.

Historical Development

Although forerunners of the day hospital component of partial hospitalization were reported in Boston, Massachusetts, and Hove, England, in the mid-1930s, the first distinct day hospital was organized by Dzhagarov (1937) in Moscow, Russia, at the First Psychiatric Hospital. This program, which began operation in 1933, was characterized as a "day infirmary for the mentally ill." The impetus for the organization of Dzhagarov's day hospital was, perhaps, a precursor of the future development of the treatment modality; the first day hospital resulted from an acute bed shortage in existing hospital facilities rather than from a theoretical or philosophical rationale proposed by the originator. Thus, the first partial hospitalization program was a financial expediency intended to serve as a substitute for the total hospitalization of severely disturbed psychiatric patients. Dzhagarov reported that 1,225 patients were treated in the program between 1933 and 1937, with an average length of stay of about 2 months. Almost half of the patients treated during this period were diagnosed as schizophrenic.

The history of the day hospital in the Western hemisphere is marked by the nearly simultaneous but independent development of programs in both Canada and England. The first organized program was established by Cameron (1947) at the Allen Memorial Institute of Psychiatry in Montreal, Canada, in 1946. It was Cameron who first introduced the term *day hospital* to describe this new treatment form. Cameron conceptualized his program as an "extension and supplement of full-time hospitalization," not necessarily as a substitute or alternative to inpatient treatment, as was the Russian program previously instituted by Dzhagarov.

The initial day hospital in England was organized by Bierer (1951) at the Institute of Social Psychiatry in London. Although conceived in early 1947, the program did not begin actual operation until early 1948. Initially, Bierer did not consider his experimental program as a replacement for the inpatient service. Rather, he states that "this new experiment, called Day Hospital, should close the gap between the inpatient department on the one hand and the outpatient department on the other" (p. 54). Organized from a psychoanalytic perspective, the London program included individual and group therapy, physical treatment, occupational and recreational therapy, psychodrama, art, and social club therapy. Bierer reported that between 1948 and 1950 about 500 adults and 150 children per year were treated in the program. Later, as the number of day hospitals increased and the form of treatment offered in these programs began to proliferate, Bierer (1962) called for a more precise definition for the term *day hospital.* Among other conditions, he proposed that a treatment unit be called a "day hospital" only when it "replaces a mental hospital." Thus, from almost the outset of its development in the Western world, the function of the day hospital (i.e., as an adjunct to or a replacement for the traditional inpatient unit) has been a focus of disagreement and debate.

In the United States, day treatment programs were in operation at the Yale University Clinic and the Menninger Clinic as early as 1948. Although the former contained both day and night components, both programs were designed as transitional treatment facilities for patients who had been hospitalized on a full-time basis. In the early 1950s, day hospitals were established in conjunction with a variety of treatment programs, including those operated by state hospitals, state departments of mental health, and the Veterans Administration hospital system. A parallel growth in day hospital programs occurred in England during this period and is thoroughly reviewed by Farndale (1961). Finally, the greatest growth in day treatment programs in the United States occurred following the passage of the Mental Retardation Facilities and Community Mental Health Centers Construction Act of 1963, which made partial hospitalization a mandated (required) service in federally funded projects.

Patterns of Expansion and Growth

It is difficult to ascertain definitive figures regarding the growth of partial hospitalization in the United States. Although several studies and surveys have been conducted in recent years, they are problematic on several counts. First, reports frequently do not describe comparable samples, thus making comparisons from period to period difficult. Second, survey data are often based on incomplete response rates. Third, surveys are frequently based on the incomplete identification of all existing partial hospitalization programs. And, finally, surveys and studies have frequently covered overlapping time periods, making comparative growth rates difficult to determine. Therefore, data reported in these surveys must be considered as approximations rather than as definitive statements. Within these limits, however, there are indications of a steadily increasing growth in both the number of partial hospitalization programs and the number of patients treated in these programs over the past 15 to 20 years.

The earliest available figures regarding partial hospitalization programs are revealed in the report of the American Psychiatric Association's 1958 conference on day hospitals. At that time, a scant eight programs were known by conference organizers to be in operation in the United States (Proceedings, 1958). Despite this small population, the conferees made several broad recommendations regarding the future development of day treatment programs based on the expectation that the modality would, indeed, become an increasingly significant force in the treatment of psychiatric patients. These recommendations proposed (1) the implementation of the day hospital concept by state departments of mental health as an "experimental substitute" for new or expanded psychiatric hospital buildings, (2) the incorporation of a day center in every mental hospital building being planned at the time, and (3) the incorporation of the day hospital experience into training programs for residents, interns, and medical students by psychiatric training centers. In light of the limited number of programs known at the time, these might be considered rather ambitious, though admirable, recommendations.

By 1963, the optimism of the American Psychiatric Association seemed, to some degree at least, to be approaching reality. In this year there were approximately 141 partial programs located in 114 mental health facilities, according to the first National Institute of Mental Health survey of such facilities (Conwell, Rosen, Hench, & Bohn, 1964). An estimated 7,689 patients were being treated in these programs, excluding those being serviced in Veterans Administration day hospital facilities. The mean case load per program, again excluding Veterans Administration hospitals, was 25. It is of interest to note that approximately one-half of these programs were located in the states of California, New York, Ohio, and Pennsylvania. Indeed, in 1963 there were 20 states that had no day treatment program at all.

During the next 5 years, as the initial impact of the Community Mental Health Centers Construction Act was being felt, continued growth in partial hospitalization occurred. By 1968, an estimated 185 partial hospitalization programs were operating in 141 facilities. This represented 139 day, 18 evening, 24 night, and 4 weekend programs. It was estimated that all partial hospitalization programs in the United States treated about 12,250 patients with an average case load of 39 per program (Glasscote, Kraft, Glassman, & Jepson, 1969). Geographically, day programs were still concentrated in five states (California, New York, Ohio, Pennsylvania, and Massachusetts), which accounted for about one-half of all reported programs. Although Community Mental Health Centers did account for over 20% of the new day hospital programs initiated from 1963 to 1967, the greatest growth during this period occurred in psychiatric hospitals (32%) and psychiatric services in general hospitals (37%). Unfortunately, data from this survey conducted by the Joint Information Service of the American Psychiatric Association and National Association for Mental Health are not directly comparable to the 1963 National Institute of Mental Health survey due to a discrepancy in the types and number of facilities surveyed in the two efforts. Nevertheless, the 1968 survey report confirmed a substantial increase in the number of partial hospitalization programs, the number of patients served, and the average case load per program. However, despite this growth, for every patient treated in partial hospitalization programs, 40 patients were treated in inpatient facilities (Glasscote *et al.*, 1969).

Partial hospitalization experienced its greatest growth between 1968 and 1972. By 1972, under the full impact of the Community Mental Health Center program, at least 989 programs were treating approximately 118,343 patients. This represented a dramatic 700% increase in the number of programs and a 14-fold increase in the number of patients served over 1963 (Taube, 1973). Community Mental Health Centers accounted for some 211 (43%) new programs from 1968 to 1972; they were providing 30% of all day care programs in the United States. The geographical distribution of programs also shifted somewhat during this period. As of 1972, 67% of all day treatment programs were located in 15 states. However, 33% were in the same four states that dominated the field in 1963; by this time, however, only one state had no programs at all. Again, despite the growth in partial hospitalization programs, Glasscote's (1969) observation was still true: For every patient treated in a partial hospitalization program, 40 were treated in inpatient services.

The latest available figures indicate continued growth in partial hospitalization programs through 1973 (Taube & Redick, 1976). By the end of that year, approximately 1,290 programs serviced 186,000 patients. This represented 3.4% of all patient care episodes in mental health facilities in 1973. This increase reflects the overall shift in location of treatment episodes during the past 20 years from predominantly inpatient settings in 1955 to predominantly outpatient settings in 1973. Day care and outpatient episodes now account for 68% of all mental health episodes, whereas only 23% were outpatient episodes in 1955.

Although absolute figures cannot be determined, available data do indicate a trend toward the expansion and growth of partial hospitalization services (see Table 1). This trend appears to be consistent in both the number of programs and the number of patients served. Coupled with the increased utilization of all forms of outpatient services, the early optimism of proponents of partial hospitalization may be a more realistic possibility. Nevertheless, it must be noted that the growth of partial hospitalization is still quite limited when compared to the growth and utilization of other psychiatric services, especially other forms of outpatient treatment.

Settings of Partial Hospitalization Programs and Diagnostic Categories Treated

Another perspective regarding the growth of partial hospitalization is obtained when one views the settings in which these programs have developed. Along with the statistical growth of day treatment programs has come a proliferation of the treatment settings for these programs as well as an expansion of the types of patients treated. This section will briefly review the growth of partial hospitalization in terms of their classification by treatment location and patient-type served.

The initial survey of partial hospitalization services (Proceedings, 1958) alluded to eight programs located in two basic types of facilites: (1)

TABLE 1. Growth in Partial Hospitalization

	Number of programs	Number of patients	Average caseload per program	Geographical distribution
1963 (Conwell et al., 1964)	141	7,689	25	½ in 4 states
1968 (Glasscote et al., 1969)	185 (500 — Taube, 1973)	12,250	39	½ in 5 states
1972 (Taube, 1973)	989	118,343	18–19 (CMH centers only)	67% in 15 states
1973 (Taube & Redick, 1976)	1,290	186,000	—	—

general psychiatric hospitals and (2) state- or county-operated hospitals. By 1973, this distribution had been complemented by several other settings. Table 2 presents data from several sources demonstrating these changes.

A close examination of this table reveals several waves of growth at different times and in a variety of settings. For example, between 1963 and 1968, partial programs experienced their growth in psychiatric hospitals and general hospital psychiatric services. The impact of the Community Mental Health Centers Construction Act of 1963 was not extensive during these initial 4 years, particularly as federally funded programs began to be implemented on a gradual basis. However, during the next 5 years the situation was altered dramatically by federal funding; Community Mental Health Centers played the predominant part in the development of day treatment programs from 1968 to 1973. It is of interest to note, also, that partial programs operated in state and county hospitals more than doubled between 1968 and 1972, as did programs in Veterans Administration hospitals during the same period.

A review of the literature also offers some indications of the expansion of partial hospitalization programs in regard to setting. Here we find reports of programs operated in state hospitals (Steinman & Hunt, 1961; Williams, 1969), Veterans Administration hospitals (Weinstein, 1960), private psychiatric hospitals (Goshen, 1956), general hospitals (Moll, 1957), Community Mental Health Centers (Silverman & Val, 1975), and even military settings (Quesnell & Martin, 1971). The military day treatment program conducted at Chanute Air Force Base in Illinois is perhaps the most unusual setting described in the literature. It was designed to provide substantial psychiatric treatment for military personnel while continuing to involve them in routine squadron activities as much as possible to avoid total withdrawal. The authors enumerate several features of the program distinguishing it from traditional day hospital programs due largely to the treatment setting. These features include (1) minimal, if any, involvement of family members in the patient's treatment, (2) a treatment goal of helping patients recognize the *existence* of psychiatric problems, and (3) the integration of the program with an inpatient service (Quesnell & Martin, 1971). In summary, it can be seen that in addition to statistical growth, partial hospitalization has experienced a diversification in treatment settings as well.

Concurrently with this diversification of settings an expansion in the types of patients treated in partial hospitalization programs has also occurred. Partial hospitalization, as a total treatment modality, has been criticized for its restrictive admission policies and limited treatment of the full range of diagnostic categories. For example, Hogarty (1971)

Table 2. Types of Facilities Reporting Partial Hospitalization Programs 1958–1973

Type of facility	1958 (Proceedings, 1958)	1963 (Conwell et al., 1964)	1968[a] (Glasscote et al., 1969)	1968[a] (Taube, 1973)	1972 (Taube, 1973)	1973 (Taube & Redick, 1976)
General psychiatric hospital	5	48[b]	16	170	72	85
General hospitals with psychiatric services	—	16	10	158	174	195
Outpatient clinics	—	26	—	36	146	242
Community mental health centers	—	4	8	84	295	400
Independent or free-standing facilities	—	13	—	—	34	43
State and county hospitals	3	—	—	—	134	139
Veterans Administration hospitals	—	—	29	—	49	62
Other	—	7	16	52	85	127
Totals	8	114	139	500	989	1290

[a]Two surveys for this period vary widely in their findings.
[b]Includes state- and county-operated facilities.

stated: "My own observation in a variety of day hospitals indicates that most patients receiving treatment are depressed married women," and, further, that "both day hospital and crisis intervention or other time limited inpatient programs tend to discriminate against the majority of schizophrenic patients" (p. 24). The accuracy of this observation is open to serious question (see Chapter 6 for additional comments). Suffice it to say here that reports concerning the diagnostic composition and treatment goals and orientations of partial hospitalization programs fail to substantiate this opinion. It is at least accurate to say that practitioners operating day treatment centers do not categorize their own programs in this limited way nor conceptualize partial hospitalization in this restricted perspective. On the contrary, practitioners have tended to view partial hospitalization programs as virtually unlimited in their treatment potential and in the scope of patients treatable in the modality. Nevertheless, a brief review of the literature will demonstrate the growth of diagnostic types treated in partial hospitalization programs.

Westlake, Levitz, and Stunkard (1974) describe one of the most unique patient populations in any partial hospitalization program. Their program is directed toward the treatment of obesity. It is a 10-week program combining behavior therapy, group therapy, and family therapy conducted on a 1-day-per-week basis. The program is conducted at the University of Pennsylvania Hospital and includes the dissemination of information about obesity, preparation of meals, group behavior therapy, and evening family participation.

The military day hospital (Quesnell & Martin, 1971) described above was aimed largely at a population of patients diagnosed as character disorders; the population reported was 88% character disorders and 12% neurotic and psychotic disorders. The mean age of patients was 19.7 years.

A third category of patients frequently treated in partial hospitalization programs includes the various neurotic disorders. Several reports describe programs designed to incorporate such patients (Carney, Ferguson, & Sheffield, 1970; Harrington & Mayer-Gross, 1959).

However, the largest single category of patients reported in the partial hospitalization literature is the schizophrenic, including chronic and acute episodes in both adults and children. One early report describes a program designed to treat severely disturbed schizophrenic children (Freeman, 1959). The approach was conceptualized as a viable alternative to total institutionalization as well as an intermediate facility that maintained the child's contact with the community.

Schizophrenic adults in both acute and chronic phases have been treated in partial hospitalization programs. Several studies report on the relative success of such treatment regimes (see Chapter 6, for a more detailed discussion). In addition, Lamb (1975) has reported on the use of day treatment in handling acute episodes in the chronic patient. He states:

> The purpose of the day hospital can be to get the patient over an acute crisis, to evaluate the chronic patient and prepare him for a rehabilitation program, to provide structure and support in a situation such as the hospitalization of a relative from whom the patient derives most of his emotional support. (p. 130)

In all instances, Lamb contends that short-term goals for patients must be set to prevent partial hospitalization from becoming a lifelong resource and that both acute and chronic patients can be treated within the same program (Lamb, 1967).

Others have shared the opinion that partial hospitalization is a viable treatment modality for both acute and chronic schizophrenic patients, although diverging opinions exist regarding the methodology and/or setting for this treatment. For example, Zwerling and Wilder (1962) state: "The one general point we wish to stress here is that the data demonstrate the Day Hospital to be a feasible modality for the treatment of both acute and chronic mental illness" (p. 184). This opinion is supported by Kris (1959, 1960) and by Wilder, Levin, and Zwerling (1966), who reemphasize that day hospital is a feasible treatment modality and generally as effective as inpatient service in the treatment of the acutely disturbed for most or all phases of hospitalization. Finally, Beigel and Feder (1970) also agree that patients with acute illness or acute exacerbation of chronic illness are amenable to day treatment programs; they maintain, however, that two distinct programs must evolve for acute and chronic patients. These programs can be conceptualized as "active vs. supportive treatment" or "day hospital vs. day care programs" (p. 1270). In addition to differences in treatment forms, Beigel and Feder suggest that the locale of day hospital and day care programs should differ. Specifically, the day hospital for the treatment of acute patients should be located near an inpatient service, while the day care program for the treatment of chronic patients should be located in the community "to emphasize its social goals and to minimize the regressive pull that the hospital exerts" (p. 1270).

The treatment of chronic schizophrenics in partial hospitalization programs is well documented in the literature. Axel (1959) recommended the establishment of day hospitals specifically for the treatment of chronic schizophrenics. Subsequently, Freeman (1962) reported on a day program in a San Francisco Veterans Administration hospital treating chronic schizophrenics with a 12-year duration of illness and two to three previous hospitalizations. In addition, Guy, Gross, Hogarty, and Dennis (1969), in a study of 137 patients treated in a day hospital program, reported improvement in the symptom constellation of communication and accessibility in schizophrenics (they were found to be less hostile, less withdrawn, less suspicious, and more cooperative). Guy *et al.* concluded that "our findings indicate that for schizophrenics with schizo-affective features, the day hospital is the treatment of choice." Somewhat later, Gootnick (1971) described the use of the day hospital in the treatment of chronic schizophrenics, concluding: "It is the psychiatric day center that can most appropriately and most economically meet the therapeutic needs of the psychotic patient who has recovered from the acute episode and requires an all-day program, two to five days a week, for a considerable period of time" (p. 120).

Finally, Hersen and Luber (1977) describe a partial hospitalization program treating chronic patients, the majority of whom were diagnosed as schizophrenics characterized by multiple hospitalizations, an erratic or nonexistent work history, and general social withdrawal or isolation.

This review illustrates that concurrent with the statistical growth of partial hospitalization programs, there has been expansion in treatment settings and a diversification of patient diagnoses treated within these programs. Although limited research findings exist to demonstrate the effectiveness of partial hospitalization programs in treating distinct diagnostic categories, theoreticians and practitioners have demonstrated a flexibility that, at least, can provide the basis for future empirical investigation. As Zwerling and Wilder (1964) stated in an earlier investigation of the efficacy of day treatment with various diagnostic categories,

> our attention has been called to what we feel to be the unusually good outcome in the cases of patients diagnosed as "paranoid schizophrenia," undifferentiated or mixed schizophrenia and psychotic depressive reactions in all diagnostic categories. (pp. 183–184)

Certainly, the time is now opportune for research regarding the influences of such factors as diagnostic category, treatment setting, and treatment orientation on the efficacy of partial hospitalization as a treatment modality.

Reasons for Growth

Having traced the growth of partial hospitalization, we will now briefly review the advantages of this treatment modality — advantages that have been a major factor in the growth of partial hospitalization programs. It will be recalled that the first recorded day hospital was organized as a practical expediency arising from a financial and spatial crisis in psychiatric facilities in the Soviet Union. It is doubtful, however, that partial hospitalization would have expanded in a literally worldwide fashion had not certain advantages given the treatment form a degree of immediate face validity.

Before describing these advantages it is necessary to mention a major development that significantly altered psychiatric services and paved the way for the development of many new treatment forms. That development was, of course, the introduction of psychopharmacological agents that promoted earlier and greater control of psychiatric symptoms (Chen, Healey, & Williams, 1969; Zwerling, 1966). This control has made it feasible, in turn, to implement a variety of treatment procedures that do not require the total hospitalization of the patient. Among these procedures is partial hospitalization.

Supported by an increased ability to control overt symptomology, the day treatment concept offered the distinct advantage of providing comprehensive treatment at a cost significantly lower than inpatient care (Glasscote et al., 1969). This economic advantage included substantially lower costs per patient per day ranging from ½ to ⅓ less than inpatient care (Farndale, 1961; Herz, Endicott, Spitzer, & Mesnikoff, 1971) as well as savings in initial capital costs to establish such programs — the latter due to the fact that much of the space and equipment requirements of a traditional inpatient unit would be unnecessary in a day treatment program (Zwerling & Wilder, 1962). In addition, it has been proposed that potential cost savings may be generated by the utilization of existing facilities and space for the operation of a partial hospitalization program. However, it must be emphasized that estimates of actual cost savings will be accurate and not misleading only if inpatient and partial programs offering *comparable services* are evaluated and, in addition, those services provided to partial hospitalization programs by other hospital facilities are considered.

Additional cost savings might also be possible due to the staffing pattern of a partial hospitalization program; such a program might eliminate one or two shifts of staff each day as well as reduce the staffing week from 7 to 5 days (Zwerling, 1966). Some of these potential savings would be less applicable, however, if a comprehensive program provided evening, night, and weekend treatment (or some combination of these) in addition to the usual daily program. Such an arrangement would necessitate the utilization of a staffing pattern more comparable to the traditional 24-hour pattern, thus increasing the cost to some degree. Nevertheless, one of the advantages of partial hospitalization that has provided impetus for its growth has been the *potential* economic saving inherent in the treatment modality.

A second possible advantage of partial hospitalization is found in the maximum flexibility in programming possible without the restrictions imposed when 24-hour patient care is required. Typically, a partial hospitalization program can be more active and varied in its therapeutic approach than an inpatient unit (Glasscote *et al.*, 1969). Community resources can be utilized in a more comprehensive and consistent manner. The treatment program itself (especially in terms of days utilized) can be tailored more adequately to the individual patient's needs and strengths (e.g., attendance can be arranged to accommodate a part-time job or other constructive activities in which the patient is engaged). And finally, activities and components of the program itself can be arranged with more cognizance of participants' assets and deficits, thus making treatment more relevant and specific (Hersen & Luber, 1977; Luber & Hersen, 1976).

Third, partial hospitalization offers the advantage of providing treatment without completely disrupting the individual's existing social system and environment. That is, the treatment setting makes it possible to lessen the "social breakdown syndrome" (Chen *et al.*, 1969) characteristic of total hospitalization. Thus, when such is indicated, the patient treated in a partial hospitalization program is able to maintain some useful ties to family and community; those social skills required for life maintenance that the patient has acquired are retained and continued utilization is encouraged; and a situation is provided in which total dependence on hospital resources and personnel is discouraged. Indeed, partial hospitalization programs provide intensive treatment possibilities that are broad in spectrum and long in duration. They can be designed to provide massive assistance around a variety of significant social problems faced by patients including vocational, educational, and recreational issues while the patient is in a position to actively participate in this assistance. In other words, partial hospitalization presents a treatment modality in which "the every-day connection to the surrounding reality is not lost" (Dzhagarov, 1937). Or, as Garner (1968) described it, all forms of partial hospitalization "have as their goals the prevention of chronic disability and institutionalization, the hastening of participation in social activities customary for the person and to encourage expansion of such activities and a constructive, creative outlook as against family separation and isolation from the patient" (p. 473).

A fourth, and related, advantage of partial hospitalization concerns the decreased amount of social stigma apparently attached to this form of psychiatric treatment (Zwerling, 1966). Although such attitudes are difficult to assess accurately, there are indications that partial hospitalization programs are more readily accepted as treatment options than inpatient services by families of patients; indeed, not only are families more accepting of this treatment alternative but they frequently act in a manner that facilitates the patient's treatment in partial hospitalization

(Zwerling & Mendelsohn, 1965; Zwerling & Wilder, 1964). Of course, there may be a variety of reasons for this action (e.g., the feeling that if inpatient hospitalization is not possible, partial hospitalization at least removes the patient from the family situation for a considerable period of time each day). Nevertheless, because partial hospitalization is less restrictive (providing relatively free access to the community and family), less disrupting to the patient's total environment, and apparently more acceptable to patients and their families (Zwerling & Wilder, 1964), it is possible to speculate that less social stigma is attached to this form of treatment than to 24-hour treatment programs. This more positive attitude, in turn, may provide a distinct advantage to partial hospitalization treatment in that the patient will be more amenable to therapeutic intervention, thus favoring a more successful treatment outcome.

Problems of Partial Hospitalization

Despite the advantages of partial hospitalization enumerated above and its gradual but steady growth in recent years, several problems are still inherent in the treatment modality. These problems will be considered in detail in several of the following chapters. Here we will attempt only to outline briefly some of the issues involved.

Almost from its inception, partial hospitalization has been plagued by the problem of terminology. That is, little consensus has been reached regarding whether this type of treatment program should be called a *day hospital* or a *day care* service and what criteria should be used in applying these terms to particular programs. A variety of opinions have been offered usually focusing on either the locale of treatment and the degree of chronicity of the patient population (Beigel & Feder, 1970) or the function of the program in terms of replacing or supplementing traditional inpatient services (Bierer, 1962). In Great Britain, both terms are currently used with the distinction determined by the types of services offered (i.e., the "day hospital" offers all treatment forms available in a psychiatric hospital while the "day care center" provides social and occupational services but only limited medical supervision).

The National Institute of Mental Health proposed a similar distinction when it identified day-night services as "a therapeutic facility for patients with mental or emotional illness or mental retardation who spend part of the day or night in a planned treatment program in this facility *and* in which a psychiatrist is present on a regularly scheduled basis who assumes medical responsibility for all patients" (Conwell *et al.*, 1964, p. 107). It appears that the term *partial hospitalization* originated in the United States as a *legal* attempt to ensure the provision of evening and weekend services in federally supported programs; the term does not allude to function but generically describes a place, a form of treatment, and/or a service with various purposes and functions (McNabola, 1975). To date, no uniform term has evolved to describe this treatment modality although the term *day hospital* has generally become less frequently used as services have expanded to include evening, night, and weekend components. Nevertheless, this inability to define terminology does create difficulties in communication. Chapter 7 will consider the issue in greater detail.

A second significant problem relates to the definition of function. As has been indicated previously, various possibilities have been postulated regarding the function of partial hospitalization. It has been suggested that partial hospitalization can serve as an alternative to full-time hospitalization, as a transitional facility for reentry into the community, as a rehabilitation setting for the chronically ill, or as a service provided to a specified population. It is highly doubtful that a single partial hospitalization program can fulfill *all* of these functions concurrently, although some combinations may be possible. Nevertheless, it is imperative that individual programs determine the *primary* task or function it wishes to fulfill and that its design be based on a clear understanding of that task (Astrachan, Flynn, Geller, & Harvey, 1970). To date, programs have had limited success in this endeavor. Frequently, subjected to administrative or treatment pressures, partial hospitalization programs attempt to satisfy divergent and sometimes contradictory needs within their treatment sphere, thus obviating their own function. As Glaser (1972) has observed, "Partial hospitalization programs have been inchoate, presenting no unified form or function to anyone. They have tried to be all things to all men and have succeeded only in not being utilized" (p. 236).

Associated with this problem are the difficulties partial hospitalization programs have demonstrated in specifying a population to be served. As indicated previously, although a variety of diagnostic categories have been treated in partial hospitalization programs, little research evidence exists to support the greater effectiveness of the approach with certain diagnoses as opposed to others. Likewise, whether partial programs should be geared primarily to acute or chronic patients (or some combination) has been the subject of much speculation but only limited supporting data. It still remains to define the parameters of treatment in regard to diagnosis and position on the acute–chronic continuum. Chapters 4, 5, and 8 will consider various aspects of the latter two problems.

Fifth, research in the area of partial hospitalization has not kept pace with its rather rapid development. Consequently, a great many important issues and questions remain unresolved and unanswered. The key problems in this area are considered in Chapter 6.

Finally, financing of partial hospitalization services has proven to be difficult in many instances. Many insurance programs, although providing coverage for inpatient and outpatient services, do not support (or support at a limited level) partial hospitalization treatment. Similarly, funds provided by federal and state programs are generally at a level significantly below the actual cost of services provided. One important and innovative attempt to generate alternative funding sources is described in Chapter 9.

Summary

This chapter has reviewed the historical development of partial hospitalization as well as it growth. The latter was discussed in terms of increases in number of programs, number of patients treated, diversification of treatment settings, and diagnostic categories treated. In addition, potential advantages of partial hospitalization were discussed and problem areas were outlined.

References

Astrachan, B. M., Flynn, H. R., Geller, J. D., & Harvey, H. Systems approach to day hospitalization. *Archives of General Psychiatry*, 1970, 22, 550–559.

Axel, M. Treatment of schizophrenia in a day hospital. *International Journal of Social Psychiatry*, 1959, 5, 174–181.

Barnes, R. Foreword. In R. Epps & L. D. Hanes (Eds.), *Day care of psychiatric patients*. Springfield, Illinois: Charles C Thomas, 1964.

Beigel, A., & Feder, S. L. Patterns of utilization in partial hospitalization. *American Journal of Psychiatry*, 1970, 126, 1267–1274.

Bierer, J. *The day hospital: An experiment in social psychiatry and syntho-analytic psychotherapy*. London: Washburn & Sons, 1951.

Bierer, J. The day hospital: Therapy in a guided democracy. *Mental Hospitals*, 1962, 13, 246–252.

Cameron, D. E. The day hospital. *Modern Hospital*, 1947, 69, 60–62.

Carney, M., Ferguson, R., & Sheffield, B. Psychiatric day hospital. *Lancet*, 1970, 1, 1218–1220.

Chen, R., Healey, J., & Williams, H. *Partial hospitalization: Problems, purposes and changing objectives*. Topeka: R. R. Sanders, 1969.

Conwell, M., Rosen, B., Hench, C., & Bohn, A. The first national survey of psychiatric day-night services. In R. Epps & L. Hanes (Eds.), *Day care of psychiatric patients*. Springfield, Illinois: Charles C Thomas, 1964.

Dzhagarov, M. Experience in organizing a day hospital for mental patients. *Nuropathologia i psikhiatria (Neuropathology and Psychiatry)*, 1937, 6, 137–147.

Farndale, J. *The day hospital movement in Great Britain*. New York: Pergamon Press, 1961.

Freeman, A M Day hospitals for severely disturbed schizophrenic children *American Journal of Psychiatry*, 1959, *115*, 893–898

Freeman, P Treatment of chronic schizophrenia in a day center *Archives of General Psychiatry*, 1962, *7*, 259–265

Garner, H Hospitalization A desirable procedure for mental illness *Comprehensive Psychiatry*, 1968, *9*, 465–473

Glaser, F The uses of the day program In H Barten & L Bellak (Eds), *Progress in community mental health* (Vol 2) New York Grune & Stratton, 1972

Glasscote, R , Kraft, A M , Glassman, S , & Jepson, W *Partial hospitalization for the mentally ill A study of programs and problems* Washington, D C The joint Information Service of the American Psychiatric Association and the National Association for Mental Health, 1969

Gootnick, I The psychiatric day center in the treatment of the chronic schizophrenic *American Journal of Psychiatry*, 1971, *128*, 485–488

Goshen, C The day hospital of the Robbins Institute *Psychotherapy*, 1956m *1*, 272–288

Guy, W , Gross, G M , Hogarty, G E , & Dennis, H A controlled evaluation of day hospital effectiveness *Archives of General Psychiatry*, 1969, *20*, 329–338

Harrington, J , & Mayer-Gross, W A day hospital for neurotics in an industrial community *Journal of Mental Science*, 1959, *105*, 224–234

Herson, M , & Luber, R F The use of group psychotherapy in a partial hospitalization service The remediation of basic skill deficits *International Journal of Group Psychotherapy*, 1977, *27*, 361–376

Herz, M I , Endicott, J , Spitzer R L , & Mesnikoff, A Day versus inpatient hospitalization A controlled study *American Journal of Psychiatry*, 1971, *10*, 1371–1382

Hogarty, G E The plight of schizophrenics in modern treatment programs *Hospital and Community Psychiatry* 1971, *22* 197–203

Kris, E Intensive short-term therapy in a day facility for control of recurrent psychotic symptoms *American Journal of Psychiatry* 1959, *115*, 1027–1028

Kris, E Intensive short-term treatment in a day care facility for the prevention of rehospitalization of patients in the community showing recurrence of psychotic symptoms *Psychiatric Quarterly* 1960, *34* 83–88

Lamb, H R Chronic psychiatric patients in the day hospital *Archives of General Psychiatry*, 1967, *17* 615–621

Lamb, H R Training long-term schizophrenic patients in the community In L Bellak & H Barten (Eds), *Progress in community mental health* (Vol 3) New York Brunner/ Mazel, 1975

Luber, R F , & Hersen, M A systematic behavioral approach to partial hospitalization programming Implications and applications *Corrective and Social Psychiatry and Journal of Behavior Technology Methods and Therapy*, 1976, *22*, 33–37

McNabola, M Partial hospitalization A national overview *Journal of the National Association of Private Psychiatric Hospitals* 1975, *7*, 10–16

Moll, A Psychiatric night treatment unit in a general hospital *American Journal of Psychiatry* 1957, *113*, 722–727

Proceedings of the 1958 day hospital conference Washington, D C American Psychiatric Association, 1958

Quesnell, J , & Martin, C A day hospital in a military psychiatric facility *Corrective Psychiatry*, 1971, *17*, 5–16

Silverman, W , & Val, E Day hospital in the context of a community mental health program *Community Mental Health Journal*, 1975, *11*, 82–90

Steinman, L , & Hunt, R A day care center in a state hospital *American Journal of Psychiatry* 1961 *117* 1109–1112

Taube, C. A. *Day care services in federally funded community mental health centers.* Statistical Note no. 96, Survey and Reports Section, Biometry Branch, National Institute of Mental Health, Rockville, Maryland, October 1973.

Taube, C. A., & Redick, R. *Provisional data on patient care episodes in mental health facilities.* Statistical Note no. 127, Survey and Reports Branch, Division of Biometry and Epidemiology, National Institute of Mental Health, Rockville, Maryland, February 1976.

Weinstein, G. Pilot programs in day care. *Mental Hospitals,* 1960, *11*, 9–11.

Westlake, R., Levitz, L., & Stunkard, A. A day hospital program for treating obesity. *Hospital and Community Psychiatry,* 1974, *29*, 609–611.

Wilder, J. F., Levin, G., & Zwerling, I. A two-year followup evaluation of acute psychiatric patients treated in a day hospital. *American Journal of Psychiatry,* 1966, *122*, 1095–1101.

Williams, J. Use of day treatment center concepts with state hospital inpatients. *American Journal of Orthopsychiatry,* 1969, *39*, 748–752.

Zwerling, I. The psychiatric day hospital. In S. Arientie (Ed.), *Handbook of Psychiatry* (Vol. 3). New York: Basic Books, 1966.

Zwerling, I., & Mendelsohn, M. Initial family reaction to day hospitalization. *Family Process,* 1965, *4*, 50–63.

Zwerling, I., & Wilder, J. F. Day hospital treatment for acute psychotic patients. In J. Masserman (Ed.), *Current psychiatric therapies.* New York: Grune & Stratton, 1962.

Zwerling, I., & Wilder, J. F. An evaluation of the applicability of the day hospital in the treatment of acutely disturbed patients. *Israel Annals of Psychiatry and Related Professions,* 1964, *2*, 162–185.

2

Behavioral Approaches to Partial Hospitalization

Thad A. Eckman, Jr.

Introduction

The Community Mental Health movement has been referred to as the third "revolution" in the history of psychiatry (Liberman, King, & De-Risi, 1976). The first was the humanization of treatment methods by Pinel, Tuke, Dix, and a limited number of other social activists of that time (Ewalt & Ewalt, 1969; Goshen, 1967). The second was the pervasive influence of psychoanalysis. Reform in the care of mental patients in the United States began more than a hundred years ago as part of a larger social movement that included prison reform, abolition of slavery, and women's suffrage (Bloom, 1975). But the wave of optimism and belief in human perfectability that was eminent at that time dimmed considerably over the next several decades. The moral-treatment movement was nearly destroyed as state mental hospitals grew in size and number while the quality of treatment deteriorated. States allocated too little money to provide adequate care for patients and there were not enough training programs to produce a sufficient number of skilled people to staff the hospitals.

At the same time, psychiatry was developing as an area of specialization in medicine and as Bloom (1975) points out, "In order for psychia-

Thad A. Eckman, Jr. • Camarillo-UCLA Neuropsychiatric Institute, and California Lutheran College, Thousand Oaks, California 91360. The preparation of this manuscript was supported in part by Grant MH26207 from the Mental Health Services Research and Development Branch of the National Institute of Mental Health.

trists to develop and maintain some respectability among their medical colleagues, they had to find organic causes and organic treatments" (p. 13). Humane treatment that emphasized a warm, understanding, supportive approach was not considered medical and psychotherapeutic approaches available today had not yet been developed, so the hospitals gradually filled up with patients who were labeled chronic and untreatable. The medical model with its orientation toward organic pathology dominated thinking about approaches to treatment for the next 80 years and remains a predominant influence today.

Shortly after World War II, the orientation toward institutional psychiatric treatment began to undergo a radical change brought about by three concurrent developments: the emergence of tranquilizing drugs (MacDonald & Tobias, 1976; Tobias & MacDonald, 1974); the development of the philosophy of the "therapeutic community," which had as a premise the notion that potential for treatment resides in patients as well as staff (Paul, 1969); and the move to locate patients in hospital units according to geographic region rather than according to diagnosis, which had resulted in growing numbers of patients being transferred to chronic treatment wards where they remained for many, many years and frequently for a lifetime (Bloom, 1975; Gilligan, 1965; Ullmann, 1967).

These developments dramatically changed approaches to treatment and caused mental health practitioners to consider alternate strategies for the optimal utilization of mental institutions within the spectrum of the larger health-care delivery system. The press for community-based programs intensified. Prolonged institutionalism was seen as undesirable and for many it was considered unacceptable (Baker, Schulberg, & O'Brien, 1969; Ozarin & Levinson, 1969). In the wake of these changes momentum for partial hospitalization in lieu of long-term inpatient care was growing rapidly. The need for other essential services was also widely recognized. Finally, in 1963, the Community Mental Health Centers Act was signed into law by President Kennedy. The act required the centers to provide five essential services: outpatient care, inpatient care, emergency services, consultation and education, and partial hospitalization. Later, five additional services were mandated. Designed to make mental health services available to communities all across the country, the legislation was also intended to provide the impetus to put newly developed psychotherapy techniques into practice.

The initiation of new social programs often brings with it an abundance of criticism, and the community-mental-health-center movement is no exception. Beginning about 1965, criticism first appeared in the published literature and increased steadily over the next several years. One of the most basic and telling criticisms has been directed toward the conceptual models of psychopathology. Controversy over the adequacy of the medical, or disease, model for treating problems associated with emotional disturbance resurfaced when it became evident that community care of mental patients would be under the administrative control of physicians. Reiff (1968) identifies several problems with the way psychiatry has historically modified the disease model, and Bloom (1975), after a review of the recent literature in this area, concludes:

> In summary, then, criticism has been leveled against the use of the disease, or medical, model to conceptualize emotional disorders, because physiological abnormality is so rarely found, because to label atypical or deviant behavior "sick" does a grave injustice to people as well as to the time-honored concept of disease, because so much of what we call psychopathology represents, not pathology in the person but pathology in the relationship between the person and the environment, and because preventive intervention must be aimed as much toward reducing sources of stress as toward increasing people's resistance to stress. (p. 214)

In addition to concerns expressed about the medical model, critics have also lamented the failure of community-based mental health services to provide alternative methods of care for the mentally ill that adequately replace the traditional treatment modes provided by the state hospitals. One of the more pervasive indictments has been tendered by the Center for Study of Responsive Law headed by consumer advocate Ralph Nader (Chu & Trotter, 1974; Chu, 1974). The report concludes with a flat statement that community mental health centers (which, of course, includes partial hospitalization programs) have largely failed to fulfill any of their major stated goals. While the report has been criticized on several grounds, particularly with respect to the information-gathering approach used (Cole, 1974; Farnsworth, 1974; Marmor, 1974), many of the claims ring true and mental health professionals would do well to follow the lead of Faberow (1973), who makes this statement in reference to the Nader Task Force report:

> To me, the report issues a clear call for change, not just in form and structure, but in substance and philosophy. It substantiates the feeling that this is a time for innovative ideas in mental health, ideas growing out of creative humanitarian and scientific concern. (p. 394)

The challenge for innovative treatment approaches to partial hospitalization requires the development of techniques and programs that can reintegrate mental patients into the community effectively and efficiently. Contemporary partial hospitalization programs, for the most part, are clearly not meeting this challenge. One important reason for this failure has been the reluctance of mental health professionals, administrators, and line staff members to give up some of the traditional treatment approaches that rely so heavily on the medical model, even when the effectiveness of these approaches has not been demonstrated.

Smith (1968) notes that mental illness is very different from physical illness in that problems experienced by the patient are not just a personal concern. He states that in addition to the patient's own inadequacies

> the social systems on which he depends have failed to sustain him — family, school, church, friendships, and the like. The task is not to cure an ailment inside his skin, but to strengthen him to the point where he can once again participate in the interactions that make up the warp and woof of life. It is also one of helping those subsystems in ways that promote the well-being and effectiveness of all people who take part in them. (p. 20)

Bloom (1975, p. 212) points out that the critics of the medical model advocate two alternate approaches to the remediation of psychopathology — an *educational* model and a *social* model. The educational model maintains that deviant behavior is learned and, with proper education, can be unlearned or replaced by nondeviant behavior. The social model envisions emotional disturbance as emerging from a social context, and the types of disorders currently brought to mental health practitioners can be ameliorated only if the social context is properly attended to. Thorough examination of these alternate approaches to treatment in general, and for partial hospitalization programming in particular, is important because, as Albee (1968) states, the sort of model that practitioners adopt "determines the kind of institutions we establish which, in turn, determines the kind of personnel we need." Albee believes that when the evidence that most neurotic and functionally psychotic behaviors indicate learned patterns is more generally accepted, the institutions created to remedy these problems will be educationally oriented in their treatment approaches.

Ideally, mental health service agencies that take on the educational-social approach to dealing with learned patterns of disturbance should place a heavy emphasis on treatment outcome and generalization to the natural environment. Services are improved to the extent that they demonstrate which interventions lead to definite improvements in the client's behavior. Approaches that focus on remediation of behavioral deficits and eradication of excessive inappropriate or undesirable behaviors and, at the same time, provide for "built-in" outcome evaluation strategies can address some of the most pointed criticisms recently directed toward mental health delivery systems. The behavior therapies offer one approach to this treatment philosophy.

MacDonald, Hedberg, and Campbell (1971) argue that a behaviorally oriented model can improve the effectiveness of the mental health system. Behavior therapy has the advantages of an approach that is deeply rooted in empiricism, makes use of an understandable language, and has clearly delineated treatment techniques that can be applied to a wide variety of problems presented by patients from a broad cross section of the community. Behavioral methods are brief, economical, and can be used by mental health workers who do not have graduate education or medical training. Liberman and his colleagues (1976) consider the advent of behavioral approaches to problems in mental health another "revolution." This may seem a strong statement when compared to the impact of Pinel, Tuke, and Dix, and to the influence of psychoanalysis; however, if one considers that the behavorial approaches to treatment not only constitute a clear break from the medical model but also incorporate the advantages of the educational and social models, as well as the added strength of well-specified strategies, then the notion that the behavior therapies represent a revolutionary approach to treatment does not seem so farfetched.

The recent growth in popularity of the behavioral orientation in mental health provides an opportunity to promote the types of procedures called for by critics of mental health programs. This may be more true for partial hospitalization programs than for any other link in the mental health services delivery system. Several important advances have been made in behavioral approaches to partial hospitalization services in the past 5 years (Eckman, 1976; Hersen & Luber, 1977; Liberman, 1973; Liberman, DeRisi, King, Eckman, & Wood, 1974; Liberman & Bryan, 1977; Liberman et al., 1976; Luber & Hersen, 1976; Spiegler 1972; Turner, 1975). Techniques currently being used include procedures for specifying individual goals, treatment plans such as the problem-oriented record, assertion training, systematic desensitization, anxiety management training, token economies, and educational workshops aimed at the remediation of the deficient necessary self-care, daily living, social, recreational, and vocational skills that enable individuals to function in society.

Behaviorally Oriented Treatment

The past 50 years have seen the accumulation of an impressive body of knowledge about the manner in which behavior is acquired and modified (see, for example, Bandura, 1969). A number of basic psychological principles that govern human behavior have been cast within the conceptual framework known as social learning theory. The application of these principles to the treatment of problems causing personal or interpersonal discomfort and distress is known as *behavior therapy*.

Many books on behavior therapy have appeared recently (Agras, 1972; Goldfried & Davison, 1976; Krumboltz & Thoresen, 1976; Lazarus, 1971, 1972; O'Leary & Wilson, 1975; Rachman & Teasdale, 1969; Rimm & Masters, 1974; Spence, Carson, & Thibaut, 1976; Wolpe, 1973). While all of these deal in one way or another with the sorts of problems patients bring to partial hospitalization programs, there are only a few references that focus primarily on institutional populations and psychiatric patients (Kazdin, 1975; Schaefer & Martin, 1969; Ullmann & Krasner, 1965) and to date no books have appeared that deal exclusively with the use of behavioral techniques in partial hospitalization. This is true in spite of the fact that behavior therapy techniques have been widely used in community mental health centers for some time and have been demonstrated to make treatment more effective and more efficient (Davison & Neale, 1974).

The goal of behaviorally oriented treatment techniques is to alter the patient's primary presenting problems rather than to directly bring about a change in personality or to treat underlying conflicts. Research indicates that maladaptive responses appear to obey the same laws of learning that govern normal behavior and are amenable to change through the application of basic principles of learning and behavior change (Bandura, 1969; Kazdin, 1975; Ullmann & Krasner, 1965).

The rationale for behavior therapy emerged from laboratory studies of classical and operant conditioning but the modern practice of behavior therapy has been most strongly influenced by studies of verbal behavior, modeling, cognitive learning, and social learning (NIMH, 1975).

Behavior therapy has become a particularly popular treatment for children's behavior problems and for managing institutional populations. It is widely accepted as the treatment of choice for most anxiety disorders. Among the wide variety of children's behavior problems that respond to manipulation of reinforcement contingencies are temper tantrums, thumb sucking, noncompliance, and head banging. Chronic mental patients have been taught a variety of social behaviors including grooming, dating, conversation skills, and how to make use of public agencies (NIMH, 1975).

Systematic desensitization has benefited patients with the distressing problems associated with anxiety reactions, enuresis, stuttering, and most types of phobic reactions (Bergin & Suinn, 1975). The technique proceeds by gradually exposing the patient to the feared situation by way of successive approximations. First, the patient is taught a systematic procedure to induce deep muscle relaxation. Then the patient is asked to imagine scenes from a hierarchy of situations that he and the therapist have constructed based on the nature and extent of the fear. The scenes, which have been ranked in order from least to most distressing, are imagined one by one while the patient remains relaxed, until they no longer evoke anxiety. The procedure is based on the theory (Wolpe, 1958) that anxiety responses can be weakened by stimulating competing responses such as relaxation or assertiveness in the presence of a real or imagined fear-producing situation. While the systematic desensitization procedure has been clearly delineated and extensively researched, the exact mechanism that produces the effect is not clearly understood. A recent review of the experimental literature (Kazdin & Wilcoxon, 1976) indicates that neither deep muscle relaxation nor imagining a progressive hierarchy of fear-producing scenes appears to be essential for the technique to be successful. Explanations of why the procedure works, however, may not be as important to practicing clinicians as the fact that it does work.

Another set of behavioral procedures that have been widely researched are the aversive therapy and punishment techniques for eliminating inappropriate or undesirable behavior. These techniques, which have been thoroughly reviewed by Rachman & Teasdale (1969), involve the administration of some type of painful or revulsive intervention such as electroshock or emetic drugs. These procedures are recommended only when other, more desirable, approaches have failed and when the therapist is well trained in the use of these techniques. Aversive therapy techniques have been most successful in situations where the problem behavior is self-destructive — head banging or self-mutilation — and, to a lesser extent, with drug addiction, alcoholism, and sexual disorders (NIMH, 1975; Rachman & Teasdale, 1969). A more useful type of punishment procedure, particularly in partial hospitalization settings where the open ward makes the tight controls necessary with the aversive techniques extremely difficult, is the removal of a privilege. Withdrawing a privilege reduces the likelihood that the effect will generalize beyond the specific behavior being punished and when the undesirable behavior has been eliminated, the patient may be more amenable to the positive aspects of his environment. The possibility of being rewarded for more appropriate behaviors increases.

While systematic desensitization and the aversive therapies have been effective for a relatively broad range of problems, these techniques are not likely to find wide use in partial hospitalization programs because the techniques require more rigorous controls and closer monitoring than can generally be achieved in an open ward where the number of patients exceeds the number of staff by a ratio of six or seven to one. These techniques are mentioned here because they illustrate

behavioral treatment interventions and because they have been successfully implemented in at least one partial hospitalization program
(Liberman & Bryan, 1977; Liberman *et al.*, 1976).

We turn now to two other behavioral approaches that have far-
reaching significance for partial hospitalization programs. Social skills
training techniques and token economies have both been demonstrated
to "help prevent or overcome the iatrogenic deterioration and social
breakdown that usually accompany prolonged institutionalization"
(NIMH, 1975, p. 326).

Training Social and Daily Living Skills

The movement away from custodial care has been supported by
evidence that the social structure of traditional custodial treatment settings and the quality of interaction between staff and patients have often
retarded the effects of treatment (Moos, 1974; Paul, 1969), resulting in
what has been termed the "social breakdown syndrome" (Zusman,
1966).

The effectiveness of day care for mental patients was thoroughly
demonstrated in a series of projects funded by the National Institute of
Mental Health (NIMH) more than 15 years ago. An NIMH report (1975)
concludes:

> It is now clear that community-based day-care and home-care programs (a)
> can be operated more economically than inpatient care; (b) have reduced
> transfers to State hospitals; and (c) can help persons who have been hos-
> pitalized make the transition to the community. (p. 365)

However, it is also clear that day care, whether in a partial hospitalization program or based in the community, does not guarantee that the
treatment of patients will be more than custodial in nature (see Rieder,
1974).

After a thorough review of the literature in psychology and
psychiatry, Paul (1969) concluded that the most important characteristics
of the inability to live successfully in the community are lack of employment, bizarre behavior, creating management problems for families or
board and care facilities, and a lack of social participation. Additionally,
the most distinguishing characteristic of people who are not helped by
hospitalization is the absence of social skills (Gripp & Magaro, 1974;
Hersen & Eisler, 1976).

The first step in any treatment program must be to see that the
patient's daily living needs are met, and the second is to assist the
patient in the acquisition and sharpening of appropriate social skills and
emotional expressiveness. We are, after all, a social animal. It is impera-

tive that partial hospitalization programs make every effort to structure treatment programs and settings in ways that afford and maximize opportunities for social skills development. Too many programs, even some of the so-called progressive programs occasion conditions that may foster apathy and social isolation.

Social skills training procedures have received wide application in recent years. Early research examined the relative importance of specific components of the techniques (instructions, modeling, behavior rehearsal, coaching, and specific feedback on performance) in a series of short-term analogue studies with nonassertive college students (McFall & Lillesand, 1971; McFall & Marston, 1970; McFall & Twentyman, 1973). A short time later analogue studies were carried out with psychiatric patients (Eisler, Hersen, & Miller, 1973; Goldstein, Martens, Hubben, van Belle, Schaaf, Wiersma, & Goedhart, 1973; Hersen, Eisler, & Miller, 1974; Hersen, Eisler, Miller, Johnson, & Pinkston, 1973).

More recently, several studies have been conducted that demonstrate the clinical application of social skills training procedures on psychiatric patients (Eisler, Hersen, & Miller, 1974; Eisler, Miller, Hersen, & Alford, 1974; Foy, Eisler, & Pinkston, 1975; Hersen & Bellack, 1976; Wallace, Teigen, Liberman, & Baker, 1973). A comprehensive review of the research literature on social skills training for psychiatric patients has been conducted by Nietzel, Winett, MacDonald, and Davidson (1977) and basic principles and applications are discussed in a chapter by Hersen and Eisler (1976).

In the past 5 years there have been two attempts to base all components of a mental health center's services on behavioral principles and methods (Liberman et al., 1976; Turner, 1975). Both of these centers, while encountering some difficulties, have experienced a great deal of success. In each case the shift from a traditional therapy program to a behaviorally oriented program was accomplished with relative ease. The experience has been illuminating because both of these researcher-clinicians and the mental health center staff members experimentally evaluated nearly every aspect of their programs. The results are very promising. Some of the more thoroughly evaluated aspects of one of the partial hospitalization programs are described in the following pages.

Personal Effectiveness Training. Liberman and his colleagues place a heavy emphasis on assisting patients to reintegrate into the community. To that end, the day treatment program was structured to provide training in the social and daily living skills that patients in partial hospitalization often lack. This group of behavior therapists have "packaged" a set of techniques aimed at helping people improve their social skills and emotional expressiveness (Liberman, King, DeRisi, & McCann, 1975). A manual and an accompanying film teach partial hospitalization therapists and other clinicians the effective use of the various components of social skills training. The authors refer to their program as *personal effectiveness* training because patients are taught how to improve the quality of their interaction in a wide variety of social, vocational, and interpersonal situations, as well as how to assert themselves appropriately. A heavy emphasis is placed on improving the expression of affect and making it consistent with the situation. Patients learn to initiate and maintain conversations, to conduct job interviews effectively, and to express sadness, anger, affection, criticism, or praise when appropriate.

The training process focuses on better use of verbal and nonverbal components of communication such as eye contact, facial expression, posture, gestures, voice quality, voice loudness, voice tone, and the content and fluency of speech. Training consists of structured, easy-to-follow procedures that bring together modeling, role playing or behavioral rehearsal, graded steps with positive feedback for small accomplishments, prompting, coaching, and the use of homework assignments to increase the likelihood that the behaviors learned in the training sessions will be effectively used in real-life situations.

Personal effectiveness training, as practiced by Liberman *et al.* (1975), proceeds in three phases: (1) the *planning meeting* where the purposes and procedures of the technique are described to patients, and individual goals are set for the second phase of the process; (2) the *training session* where problematic, real-life situations are simulated and appropriate behavior is rehearsed by the clients while the therapist prompts, models, coaches, and gives the patient feedback on his performance; and (3) the *evaluation meeting* where the therapists who conducted the training session provide one another with mutual supervision and feedback for improving their skills in the training process.

In the partial hospitalization program, the training is generally carried out in small groups of 8 to 10 patients with two staff members as leaders. Sessions are conducted twice weekly for 2 hours. The groups are open-ended in the sense that new patients are admitted to the group as they enter the day treatment program and leave the group when their individual goals have been attained. The average term of treatment is 6 to 12 weeks.

During the planning session patients are asked to report on how well they followed through with the homework assigned in the previous session. Completed assignments are reinforced with praise, respect, and interest from the therapists and applause and compliments from the group members. The report serves as the basis for targeting goals to work on in the training session that follows. If the patient successfully completed the homework assignment, a more demanding goal is selected. If the patient was unsuccessful, the goal would be to practice the situation again during the training session in order to increase the skill level and/or to reduce the anxiety associated with executing the assignment. If it appears that the assignment was too ambitious, it would be exchanged for one less difficult. The planning phase is ended when all patients have reported and received feedback on their assignments.

The process of personal effectiveness training consists of guiding the patient systematically through a series of explicitly delineated steps. First, problems that the person has in communicating and expressing feelings are identified by helping the person and his family or significant others specify where, how, when, what, and with whom the problem occurs. Next, the problem situation is simulated in a "dry run." This provides the therapist with the opportunity to conduct an analysis of the patient's verbal and nonverbal component communication skills and permits the patient and the therapist to try out various strategies for improving old behaviors and training new ones. The technique is powerful because it allows the therapist to examine the client's performance directly and, even though the role-play situation is admittedly artificial, it has been observed that the behavioral excesses and deficits generally exhibited by the patient in the actual situation are usually displayed during the role-play. The performance is then broken into specific, concrete steps so that each segment can be practiced under the careful guidance of the therapist. Explicit instructions are provided as the therapist or another client models an alternative. Components of the whole performance — eye contact, facial expression, hand gestures, voice tone, and speech content — are added one by one as the therapist gradually modifies the patient's performance, shaping each behavior until all of the excesses and deficits have been dealt with. For example, the therapist might say, "Your eye contact was very good that time and taking her by the hand really helped get her attention. This time try to soften your voice tone and let's see if that makes you sound even more sincere." The patient is then guided through another rehearsal. Each patient engages in 5 to 10 minutes of role playing. The scenes are kept short, usually lasting 1 or 2 minutes, and specific feedback is given by the therapist and other group members after each practice trial. During the training process positive feedback for improvement rather than chastisement for failure is emphasized. Little or no critical feedback is given. At the end of each session the patient is asked to complete an assignment intended to provide an opportunity to practice, in an actual situation, the improved response he has learned in the training session.

Educational Workshops. As lofty a goal as the eradication of social skills deficiencies may be, it is not enough. Atthowe (1973) has criticized programs that facilitate adjustment within the treatment setting but fail to promote prosocial behaviors necessary for survival in noninstitutional environments. Behaviorally oriented programs have not been exempt from criticism on these grounds (Bornstein, Bugge, & Davol, 1975). It seems clear that we cannot simply expect that behavior as complex as human social interaction will automatically generalize from the relatively rigid and artificial confines of the treatment setting to the countless circumstances arising from daily activities. Therapists must begin to *program* for generalization if they want to ensure that the skills their patients acquire in therapy will be used outside the treatment setting.

One way to accomplish this, of course, is to create homework assignments consistent with the patient's treatment goals and encourage the patient to carry them out. An additional, and far more effective, strategy is to create a treatment environment that makes it possible for therapists to seize upon every possible opportunity to structure activities in a manner that stretches the boundaries of that environment beyond the walls of the clinic — to structure activities in a way that makes it legitimate for therapists to conduct treatment in the community. The partial hospitalization program described by Liberman and his associates (1974, 1976) exemplifies such an approach.

Prompted by a study that indicated that fewer than 30% of the behaviors exhibited by their patients in the mental health center's day treatment program involved any kind of social participation, Liberman and his colleagues increased the type and amount of structured group activities that were likely to evoke adaptive social behaviors from patients. A number of *educational* workshops were created to remediate the social and daily living skills deficits of people attending the day care program. The workshops covered a wide variety of subject areas including conversation skills, personal finance, consumerism, grooming, current events, ethnic exchange, public agencies, vocational preparedness, anxiety and depression management, weight control, and recreation-education-social-transportation. Highly structured activities were developed and tailored to fit behavioral objectives specified for each workshop session.

Each workshop consisted of eight weekly sessions. Because the session activities of a given workshop were independent of one another, patients could be admitted to a workshop at any time and terminate involvement when their treatment objectives had been attained. So, while the educational workshop program operated continuously on an 8-week "semester" basis, patients attended only those workshops, and certain selected sessions from other workshops, that offered the greatest assistance in helping them meet their individual treatment goals. Day care staff members rotated through each workshop, first as a co-leader, then as a leader, so that every staff member could become proficient at leading each workshop. This procedure also helped to alleviate boredom among the staff members. The educational workshop constituted a curriculum of course offerings, so it was decided to refer to the patients as "students" and therapists as "counselors" in an effort to reduce the social stigma that frequently surrounds a mental health center.

Frequent program evaluations were conducted by the workshop leaders and the partial hospitalization administrator to assess all aspects of the program including objectives, materials, activities, methods, and treatment outcome. Shortly after the educational workshop program was under way, social participation among patients increased by nearly 40% over the amount present before the program was implemented. Nonsocial behaviors decreased by more than 44%. In addition, the amount of the day care center's available hours structured with some type of prosocial activity increased from 47% to 84%. Figure 1 indicates a typical weekly schedule at the Oxnard day treatment center.

The primary goal of every one of the workshops is to maximize available opportunities for patients to learn the necessary self-help skills that will permit effective integration in the community and, at the same time, provide the necessary skills and information for dealing with

	MONDAY		TUESDAY		WEDNESDAY	THURSDAY		FRIDAY	
8:30									
9:00	Program Review		Program Review		Program Review	Program Review		Program Review	
9:30	Week-End Review		Grooming Workshop		Consumerism Workshop	Personal Finances Workshop		Open	
10:00									
10:30			Personal Effectiveness Training Workshop	Conversational Skills Workshop				Personal Effectiveness Training Workshop	Conversational Skills Workshop
11:00	Current Events Workshop	Cooking Skills Workshop			Consumerism Lab Shopping	Ethnic Exchange Workshop			
11:30									
12:00									
12:30	Lunch		Lunch		No Lunch (Close Day Treatment Center at 12:30)	Lunch		Lunch	
1:00									
1:30	Medical and Clinical Consultation		Patient/Staff Meeting			Credit Incentive Program Review		Occupational Therapy Workshop	Anxiety Management Training
2:00						Medical and Clinical Consult.			
2:30	Public Agencies Workshop		Diet Workshop			Vocational Preparedness Workshop	R.E.S.T. Workshop		
3:00					Staff Training				
3:30			Vocational Preparedness Workshop						
4:00	Staff Meeting							Open	
4:30									
5:00			Open			Team Meeting			
5:30									

Figure 1. A typical weekly schedule for the partial hospitalization program conducted by Liberman and his associates.

problems in daily living. Brief descriptions of some of the more popular workshops conducted by Liberman and his colleagues follow.

Grooming. In the grooming workshop, patients are taught a variety of techniques for effective personal grooming, health care habits, and dress. Instructional techniques involve group discussion, demonstrations by a cosmetologist, a barber, and a public health nurse, as well as the workshop leader, and practice sessions where the demonstrated techniques can be applied. The goals of the grooming workshop are to provide instruction in the basic techniques of personal grooming and hygiene including bathing, dental care, and proper care of hair, skin, and nails. Patients are also taught how to apply cosmetics, how to care for clothing, and how to dress for special occasions.

Conversation Skills. The conversation skills workshop assists patients in developing verbal facility in a variety of social settings. It is especially suited for persons who have never learned appropriate verbal skills or who have not used them in a long time. The workshop sessions provide instruction in the basic skills necessary to initiate conversation, maintain a conversation that is pleasant and personally rewarding, and terminate conversations in a manner that is socially acceptable. The workshop aims to reduce speech problems and other habits that may interfere with effective interpersonal communication.

The conversation skills workshop is designed to accommodate lower functioning patients who have difficulties verbalizing in social situations. Referrals to the group are made on the basis of the counselor's observation of specific behaviors such as low rate of verbal interaction, lack of spontaneity in groups, failure to elaborate topical content in the context of conversation, and excessively changing the topic of discussion.

Workshop session activities center around a variety of situations that require a great deal of verbalization from the participants, such as conducting "man on the street"-type interviews, show and tell exercises, games that necessitate a high rate of verbal responding, and so on. Because the participants are lower functioning individuals, a great deal of modeling and behavioral rehearsal takes place. This allows the workshop leaders and other participants ample opportunity for feedback and social reinforcement. Audio and video devices are used to instruct patients directly in techniques of giving and receiving feedback.

Personal Finances and Consumerism. This workshop teaches the patient how to manage his finances and ways to obtain the most in quality and quantity of goods and services within the limits of his individual income. Patients regularly go into the community to try newly acquired skills in local supermarkets, rental agencies, and clothing stores, where they do comparative shopping. The goals of the workshop are to teach patients how to count money and make correct change, how to open and maintain a bank account, how and where to apply for credit, how to set up a system for personal record keeping, and how to plan a budget.

A unique feature of the personal finances and consumerism workshop is the integration of a token economy into the framework of the course. Many of the patients are able to use the token system to learn the basics of money management. Parallels are continually drawn between the token economy in the partial hospitalization program and "real world" finances in an effort to enhance the prospects for generalization when incomes for these people become a reality.

Recreation-Education-Social-Transportation (REST) Workshop. The REST workshop helps patients to become familiar with the recreational, educational, and social opportunities in the surrounding community. They learn how to use local transportation systems, including the use of city bus lines, Greyhound bus systems, and taxicabs. Information is provided regarding the necessary steps involved in obtaining a driver's permit and how to prepare for the driver's test. Rules of the road and safety factors are emphasized. Patients learn how to obtain a library card and how to use the public library. They learn how to plan for their own recreational activities such as picnics, camping trips, and parties. Patients learn how to go about joining social clubs and community craft classes. They learn about community college offerings, business and technical schools, and adult evening programs in the local high schools.

Class assignments require patients to gather information over the telephone and in the community at specified facilities such as the local welfare agency, a high school or college registrar's office, or at the office of parks and recreation.

Public Agencies Workshop. In the public agencies workshop, patients learn how to make appropriate use of the many public, charitable, fraternal, and political agencies in the county. The goals of the workshop are to acquaint the patients with the eligibility requirements, specific services, and location of the agencies. Patients learn how to identify problems and choose appropriate community resources for assistance. Patients are taught how to make effective phone calls and personal appearances in contacts with community agencies. They learn how to overcome roadblocks and deterrents in obtaining necessary and desired services. The topics covered in this workshop include registering complaints and grievances, rectifying an error, being persistent without being obnoxious, following up initial contacts, and anticipating lengthy waits. The primary instructional techniques in this workshop are modeling, behavioral rehearsal, and practice in the public agencies. Group discussions and short lectures are also used.

Current Events Workshop. The current events workshop is designed to familiarize patients with the various types of news media and to encourage them to become informed about local and national news.

Patients study local newspapers, national news magazines, and radio and television newscasts. Participants are asked to bring news articles to class for discussion and interpretation. Each patient keeps a notebook containing one article relating to the assigned topic for each day of the week. Patients also learn about the organization and operation of governing bodies, including the federal government, city and county council meetings, planning meetings, and zoning board meetings. Each class member is expected to attend at least one city council meeting.

The primary instructional technique in the current events workshop is group discussion, but a strong emphasis is placed on community involvement in civic affairs. Class sessions are held at the county court house, a local newspaper facility, and other community settings that relate directly to the topic at hand. One entire session is devoted to learning to write a letter to the editor of a newspaper or magazine, and how and where to write to governmental officials.

Workshop in Vocational Preparedness. The vocational preparedness workshop provides instruction and counseling in the necessary skills for securing and maintaining employment. Patients learn the basic components of a résumé and how to write a personal information record to be used when seeking employment. They learn how to obtain and complete an application for employment, and how to prepare for and take a job interview. Basic strategies for interacting successfully with fellow employees and supervisory staff are taught. Patients are instructed in appropriate methods for dealing with grievances, and how to ask for a pay raise and promotion.

The primary modes of instruction center around modeling, behavior rehearsal, prompting, structured feedback, discussion sessions, and practice in real-life situations. Employers and personnel officials are invited to give presentations, conduct interviews, and join the discussion.

Ethnic Exchange. The ethnic exchange workshop was initially instituted because the partial hospitalization program is located in a city which has a multiethnic community. The purpose of this workshop is to clarify the problem of biculturalism faced by Mexican-Americans and other cultural minorities, to identify existing forms of overt and covert discrimination, and to provide a mechanism for overcoming these problems. Topics of discussion include the histories of minority cultures, sources of racial prejudice, problems associated with discrimination, food, music, people, and strategies for solving problems. Because of the richness of community resources related to this workshop, class sessions are frequently held outside of the day treatment center. This workshop is led jointly by three staff members, one is black, one is Anglo, and one is Chicano.

Exercise Workshop. The exercise workshop is designed to help patients create an individualized exercise program based on their own particular needs. An assessment is made of those aspects of physical fitness that need the most improvement. Programs are created which involve activities that the patient enjoys. These activities include calisthenics, jogging, bicycling, swimming, tennis, and so on. Patients are encouraged to engage in these activities with friends and family to increase the opportunity for positive social interaction. In addition to the exercise program, patients are taught the rules and scoring procedures for a variety of social games such as bowling, volleyball, baseball, and many others. During the exercise workshop each patient is given an opportunity to put his personalized exercise program into effect.

Other Workshops. A variety of other workshops are taught periodically. The diet workshop teaches appropriate methods of weight control and nutrition. The self-management workshop focuses on procedures for controlling undesirable thoughts and feelings by increasing productive thinking and decreasing anxiety, and the dating workshop encourages social participation and facilitates the learning of appropriate social responses in a variety of situations, particularly the dating relationship and dating activities.

A sample session outline from a workshop leader's guide is displayed in Figure 2.

Evaluation of the Workshops. An assortment of evaluation techniques have been used to assess the progress of the educational workshop program. Nearly all of the workshop leaders in Liberman's group have made use of paper-and-pencil tests when an assessment of verbal content was needed. Records of the assessments were kept and reviewed on an individual basis with each patient's assigned therapist so that achievement of individual goals could be recorded. Some workshops, such as the conversation skills workshop, made use of a more sophisticated evaluation scheme, i.e., pre/post analysis of direct observation of patient behaviors by independent observers.

Among the more useful of the evaluation techniques is an assessment of four indices that have direct relevance to the general goals of the workshops. These are *attendance* at the workshops, *verbal participation* during the workshops, completion of *classroom assignments*, and completion of *homework assignments*. Measures of these four indices were created to provide an objective indication of the overall impact of the educational workshop program (Eckman, 1974). First, it was recognized that regular attendance at the workshops would be necessary to maximize the benefit of the learning experience. Accordingly, a heavy emphasis is placed on the importance of regular attendance. Second, an

SESSION FOUR

Staff: Leader. co-leader and guest speaker (employer from community)
Facilities. Conference Room
Equipment and Materials Desk and two chairs for interview setting, and personal interview
                 information forms for each participant.
Admission to Workshop Session 5 credits
Goal. To reach appropriate strategies for carrying out a job interview and to provide an oppor-
       tunity for each person to practice interviewing for a job

ACTIVITIES

1 Guest Speaker
     a The invited guest will discuss important personal characteristics for which he looks when
        interviewing a prospective employee
         (1) The employer will discuss why he asks specific types of questions during an interview
             and the manner in which he evaluates prospective employees.
     b The speaker's discussion will be followed by a question and answer session. (The workshop
        leader and co-leader should assist the participants in developing a relevant line of ques-
        tions)
2 Evaluate the dress of each participant emphasizing appropriate styles and types of attire. All
    participants should engage in the feedback process
3. Behavioral Rehearsal
     a Ask each participant, in turn, to role play an interview situation. The leader and co-leader
        will participate as the interviewer The interviews should be short (approximately 2 to 3
        minutes)
         (1) Give the participant immediate feedback on his performance. Encourage group
             members to participate in the feedback session Emphasize positive feedback and try
             not to dwell on negative aspects of the interview
         (2) The workshop leaders should model appropriate behaviors and interview strategies in
             an effort to shape approximations to the desired response. Other group members may
             also be called on to assist in modeling interview technique.
         (3) During the behavioral rehearsal, the workshop leaders should continue to provide
             feedback while reminding participants of the most common problem areas which are
             likely to occur during an interview
4 Following up the interview
     a Engage the participants in a group discussion regarding appropriate follow up techniques.
        The discussion should include
         (1) When to follow up
         (2) Type and frequency of follow up procedures
             (a) phone calls
             (b) letting the interviewer know how and where to reach you
5. Review the first three workshop sessions with the participants and obtain feedback from them
    regarding workshop activities and the helpfulness of the activities.

HOMEWORK ASSIGNMENT

Ask each client to approach a prospective employer in the community. He should ask a shop
manager about the availability of a job and then write the following points of information on a
card to be brought to the next session
     1 The name of the establishment where the inquiry was made.
     2 The name of the person he talked with about a job possibility
     3 Whether or not he felt comfortable during the encounter

Figure 2. A sample session outline from the vocational preparedness workshop.

effort is made to promote prosocial behavior among the students. Every workshop has this as a basic goal. Spontaneous verbal behavior is used as an indication of this index, and voluntary participation in workshop sessions is the specific measure employed. Finally, the importance of programming the generalization of learned behaviors is stressed. The method used to promote generalization is homework assignments that parallel the course content for each workshop session. The ratio of completed homework and classroom assignments to the number of assignments made provides a measure of the patient's utilization of opportunities for generalized learning.

Keeping accurate records of these four indices provides a means of making comparisons within each workshop on a session-by-session basis, across each workshop from one semester to the next, and across all workshops for a given semester.

The Credit-Incentive System. Many of the patients who attend partial hospitalization programs have become apathetic and withdrawn. Sometimes this is symptomatic of the distress the person is experiencing or the result of the social breakdown that often accompanies periods of hospitalization. Additionally, many patients come to the day care center expecting to *receive* treatment rather than to actively participate in the treatment process. Staff members tend to react to these patients with encouragement and support at first, and then follow up with continuous prompts that deteriorate to nagging when patients fail to respond. In time, these patients extinguish the therapist's attempt to involve them in program activities and the social breakdown process continues.

One approach taken by behavior therapists for motivating patients who refrain from active involvement in treatment programs is the *token economy* (Ayllon & Azrin, 1968; Paul, 1969). The efficacy of this approach has been demonstrated in a wide variety of settings ranging from inpatient facilities for chronic mental patients to public school classrooms (Davison, 1969; Gelfand, Gelfand, & Dobson, 1967; Kazdin, 1973; Kazdin & Bootzin, 1972; Patterson, 1976; Welch, & Gist, 1974).

A token economy is a structured system for rewarding individuals for performing behaviors that are considered to be appropriate or desirable. The first step in establishing a token economy in a mental health program is to create a list of *observable* behaviors that are functional and likely to be useful to the patient outside, as well as inside, the treatment setting. The second step is to create a list of rewards that the patients value. Rewards can be tangible items, activities, or social reinforcements. Finally, there must be a medium of exchange—the token. Patients can earn tokens for engaging in constructive activities and then use the tokens to purchase rewards.

A modified token economy has been successfully implemented and experimentally evaluated in a partial hospitalization program at a mental health center in Oxnard, California (Liberman, Fearn, DeRisi, Roberts, & Carmona, 1977). In this program, patients use credit cards rather than tokens. The credit card (see Figure 3) has numbers printed on it that parallel monetary denominations. The credit card is made of a piece of 4 × 3-inch cardboard. Staff members punch holes in the card when credits are earned and spent. A small die-cast punch with a symbol such as a heart, spade, or club is punched when credits are earned. The same number is overpunched by a larger, circular die when credits are spent. An advantage of the credit card over the use of tokens (Aitchison, 1972) is that patients are required to learn to manage a small amount of credit — a functional behavior in today's credit-oriented society. A second advantage is that the credit cards can be used as a permanent record of the patient's earnings and spendings. "Credit flow" can be used as an unobtrusive measure of a patient's involvement in program activities.

In the program described by Liberman *et al.* (1976), the credit-incentive system serves as a central integrating mechanism around which other program activities revolve. For example, all of the educational workshops described above require a small tuition payment because treatment is considered a valuable experience. Once the patient gains admittance to the workshop there are many ways to earn new credits. Participating in a behavioral rehearsal, modeling for another patient, and completing classroom assignments and homework assignments are just a few activities for which credits can be earned. Credits can be earned for doing day center maintenance tasks and housekeeping chores such as cooking, serving meals, dusting furniture, cleaning the stove and refrigerator, and tidying up the living room. These tasks, however, all require work inside the center so they generally do not earn as many credits as activities performed in the community, such as shopping for groceries, visiting the library, or obtaining a job application.

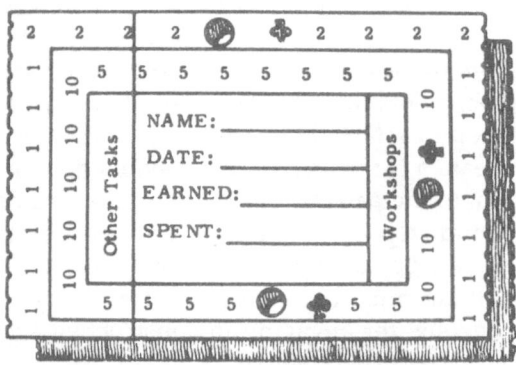

Figure 3. Credit card used by patients participating in credit-incentive program.

Credits can be spent for coffee, lunch, private staff time, listening to the stereo, a day off from day treatment, and so on.

Once each week a meeting is held where patients and staff negotiate the credit value for the lists of privileges and responsibilities existing in the partial hospitalization program. The number of credits attached to specific activities varies as a function of supply and demand. Features inherent in the token economy parallel those of the "free marketplace." The system serves as a model that helps patients in the personal finances workshop assimilate basic principles. It also has flexibility in that contingencies can be tailored to meet the specific treatment goals of individual patients. For example, a patient who is usually passive may earn credits for voluntary participation during a workshop session, while another patient who is frequently loud and obnoxious may earn credits for specified periods of "quiet time" during the same session.

The credit-incentive system at the Oxnard mental health center has had a positive impact on the staff as well as the patients. Punching the credit cards provides a vehicle for increased staff–patient contact. Social reward almost always accompanies the punching of a patient's card. The dropout rate from the partial hospitalization program has significantly decreased, as have nagging and cajoling by the staff.

The credit-incentive system is a partial solution to some of the problems associated with the lack of desired involvement from patients in day treatment programs. The technique appears to work reasonably well when the awarding of credits is paired consistently with an abundance of verbal and nonverbal praise and approval, and when the credits are dispensed immediately.

Summary

Behavioral approaches offer a viable treatment alternative for many partial hospitalization programs. The techniques of personal effectiveness training, credit-incentive systems, and the educational workshops model to teach social and daily living skills appear to hold great promise. These interventions have the advantage of a set of clearly specified treatment techniques that can be used by mental health workers from a variety of professional orientations as well as those who do not have medical training or graduate education. Behavioral techniques can generally be integrated into ongoing treatment programs with relative ease, are usually brief, and can be applied to many of the problems presented by patients representative of a broad range of socioeconomic class, educational level, ethnic background, and age.

Behavioral procedures have been implemented in partial hospitalization programs and have withstood the test of rigorous experimental evaluation. While there is little doubt that these approaches can be successfully applied, the limits of the techniques must be recognized, too. There are still very few studies that have investigated the effects of behavioral interventions for patients who were not selected for some problem considered amenable to the behavioral approach. Studies that adequately examine the durability of behavior change are only beginning to find their way into the literature. Strategies for promoting generalization of newly acquired skills outside of the treatment setting are still not well understood. The behavior therapists have achieved many successes but they have encountered some difficulties too. After several years of experience in applying behavioral approaches to partial hospitalization programs, Liberman and his colleagues (1976) concluded:

> It is our view that behavior modifiers, faced with cost accountability in community mental health centers will have to temper their characteristic optimistic and idealistic strategy of trying to produce functional and generalized behavior change in severely deficient, chronic, and relapsing psychotic individuals. (p. 596)

For such patients, whose realistically expected treatment outcome may still fall short of the outside limits of available treatment methods, it may not be enough to provide innovative programs within the mental health agency. These individuals may never be able to live autonomously in the community as it now exists. Rather than resort to the state hospital or the day treatment center as the only solution, it may be possible to reengineer certain aspects of the community to make it capable of supporting these marginally functional individuals.

We may be nearing the limits of treatment approaches available at the present state of the art. Barring some unforeseen breakthrough, the next major step in mental health services delivery may be in taking the lead to establish the necessary linkages with other community agencies that provide supportive services to those individuals who lack the skills and resources to live successfully in our society. The inherent structure and specificity of the behavioral approach make it a viable mechanism for providing continuity of care across agencies.

Finally, the current press for accountability from all agencies expending public funds requires that programs demonstrate responsiveness to the needs of the public, be efficient and inexpensive, and have an inherent capacity for outcome evaluation. The behavioral alternative, whatever its shortcomings, perhaps comes closer to meeting these demands than any other treatment model currently in use. For this, if for no other reason, it is especially deserving of serious attention.

ACKNOWLEDGMENTS

The author is grateful to Jeff Rosenstein, Dr. Tim Kuehnel, Jon Backstrom, and Holly Jaacks for their assistance. A special acknowledgment must be made to Dr. Larry King, Dr. Robert Liberman, and Dr. William DeRisi for giving impetus to mental health's fourth "revolution." The views expressed here are those of the author and do not necessarily reflect the official policy of the NIMH, the Regents of the University of California, or the Regents of California Lutheran College.

References

Agras, W. S. *Behavior modification: Principles and clinical applications.* Boston: Little, Brown, 1972.

Aitchison, R. A. A low cost, rapid delivery point system with "automatic" recording. *Journal of Applied Behavior Analysis,* 1972, *5,* 527–528.

Albee, G. W. Models, myths and manpower. *Mental Hygiene,* 1968, *52,* 168–180.

Atthowe, J. M. Token economies come of age. *Behavior Therapy,* 1973, *4,* 646–654.

Ayllon, T., & Azrin, N. H. *The token economy.* New York: Appleton-Century-Crofts, 1968.

Baker, F., Schulberg, H. C., & O'Brien, G. M. The changing mental hospital — its perceived image and contact with the community. *Mental Hygiene,* 1969, *53,* 237–244.

Bandura, A. *Principles of behavior modification.* New York: Holt, Rinehart & Winston, 1969.

Bergin, A. E., & Suinn, R. M. Individual psychotherapy and behavior therapy. In M. R. Rosenzweig & L. W. Porter (Eds.), *Annual review of psychology,* 1975, *26,* 509–556.

Bloom, B. L. *Community mental health, a general introduction.* Belmont, California: Brooks/Cole, 1975.

Bornstein, P., Bugge, I., & Davol, G. Good principle, wrong target — an extension of "Token economies come of age." *Behavior Therapy,* 1975, *6,* 63–67.

Chu, F. D. The Nader report: One author's perspective. *American Journal of Psychiatry,* 1974, *131,* 775–779.

Chu, F. D., & Trotter, S. *The madness establishment.* New York: Grossman, 1974.

Cole, J. O. Comment. *American Journal of Psychiatry,* 1974, *131,* 781–782.

Davison, G. C. Appraisal of behavior modification techniques with adults in institutional settings. In C. M. Franks (Ed.), *Behavior therapy: Appraisal and status.* New York: McGraw-Hill, 1969.

Davison, G. C., & Neale, J. M. *Abnormal psychology: An experimental clinical approach.* New York: Wiley, 1974.

Eckman, T. A. *The educational workshop model as a treatment alternative in a partial hospitalization program.* Unpublished manuscript available from the author, 1974.

Eckman, T. A. An educational workshop in conversation skills. In J. D. Krumboltz & C. E. Thoresen (Eds.), *Counseling methods.* New York: Holt, Rinehart & Winston, 1976.

Eisler, R. M., Hersen, M., & Miller, P. M. Effects of modeling on components of assertive behavior. *Journal of Behavior Therapy and Experimental Psychiatry,* 1973, *4,* 1–6.

Eisler, R. M., Hersen, M., & Miller, P. M. Shaping components of assertiveness with instructions and feedback. *American Journal of Psychiatry,* 1974, *131,* 1344–1347.

Eisler, R. M., Miller, P. M., Hersen, M., & Alford, H. Effects of assertive training on marital interaction. *Archives of General Psychiatry,* 1974, *30,* 643–649.

Ewalt, J. R., & Ewalt, P. L. History of the community psychiatry movement. *American Journal of Psychiatry*, 1969, *126*, 43–52.

Faberow, N. L. The crisis is chronic. *American Psychologist*, 1973, *28*, 5, 388–394.

Farnsworth, D. L. Comment. *American Journal of Psychiatry*, 1974, *131*, 779.

Foy, D. W., Eisler, R. M., & Pinkston, S. G. Modeled assertion in a case of explosive rage. *Journal of Behavior Therapy and Experimental Psychiatry*, 1975, *6*, 135–137.

Gelfand, D. M., Gelfand, S., & Dobson, W. Unprogrammed reinforcement of patients' behavior in a mental hospital. *Behaviour Research and Therapy*, 1967, *5*, 201–207.

Gilligan, J. Review of the literature. In M. Greenblatt, M. H. Soloman, A. S. Evans, & G. W. Brooks (Eds.), *Drug and social therapy in chronic schizophrenia*. Springfield, Illinois: Charles C Thomas, 1965.

Goldfried, M. R., & Davison, G. C. *Clinical behavior therapy*. New York: Holt, Rinehart & Winston, 1976.

Goldstein, A. P., Martens, J., Hubben, J., van Belle, H. A., Schaaf, W., Wiersma, H., & Goedhart, A. The use of modeling to increase independent behavior. *Behaviour Research and Therapy*, 1973, *11*, 31–42.

Goshen, C. E. *Documentary history of psychiatry: A source book on historical principles*. New York: Philosophical Library, 1967.

Gripp, R. F., & Magaro, P. A. The token economy program in the psychiatric hospital: A review and analysis. *Behaviour Research and Therapy*, 1974, *12*, 205–228.

Hersen, M., & Bellack, A. S. A multiple-baseline analysis of social-skills training in chronic schizophrenics. *Journal of Applied Behavior Analysis*, 1976, *9*, 239–245.

Hersen, M., & Eisler, R. M. Social skills training. In W. E. Craighead, A. E. Kazdin, & M. J. Mahoney (Eds.), *Behavior modification: Principles, issues, and applications*. Boston: Houghton Mifflin, 1976.

Hersen, M., Eisler, R. M., & Miller, P. M. An experimental analysis of generalization in assertive training. *Behaviour Research and Therapy*, 1974, *12*, 295–310.

Hersen, M., Eisler, R. M., Miller, P. M., Johnson, M. B., & Pinkston, S. G. Effects of practice, instructions, and modeling on components of assertive behavior. *Behaviour Research and Therapy*, 1973, *11*, 443–451.

Hersen, M., & Luber, R. F. Use of group psychotherapy in a partial hospitalization service: The remediation of basic skill deficits *International Journal of Group Psychotherapy*, 1977, *27*, 361–376.

Kazdin, A. E. The failure of some patients to respond to token programs. *Journal of Behavior Therapy and Experimental Psychiatry*, 1973, *4*, 7–14.

Kazdin, A. E. *Behavior modification in applied settings*. Homewood, Illinois: Dorsey Press, 1975.

Kazdin, A. E., & Bootzin, R. R. The token economy: An evaluative review. *Journal of Applied Behavior Analysis*, 1972, *5*, 343–372.

Kazdin, A. E. & Wilcoxon, L. A. Systematic desensitization and nonspecific treatment effects: A methodological evaluation. *Psychological Bulletin*, 1976, *83*(5), 729–758.

Krumboltz, J. D., & Thoresen, C. E. *Counseling methods*. New York: Holt, Rinehart & Winston, 1976.

Lazarus, A. A. *Behavior therapy and beyond*. New York: McGraw-Hill, 1971.

Lazarus, A. A. *Clinical behavior therapy*. New York: Brunner/Mazel, 1972.

Liberman, R. P. Applying behavioral techniques in a community mental health center. In R. Rubin, J. P. Brady, & J. Henderson (Eds.), *Advances in behavior therapy* (Vol. 4). New York: Academic Press, 1973.

Liberman, R. P., & Bryan, E. Behavior therapy in a community mental health center. *American Journal of Psychiatry*, 1977, *134*(4), 401–406.

Liberman, R. P., DeRisi, W. J., King, L. W., Eckman, T. A., & Wood, D. Behavioral measurement in a community mental health center. In P. O. Davidson, F. W. Clark, &

L. A. Hammerlynck (Eds.), *Evaluation of behavioral programs in community, residential, and school settings.* Champaign, Illinois: Research Press, 1974.

Liberman, R. P., King, L. W., DeRisi, W. J., & McCann, M. *Personal effectiveness: Guiding people to assert themselves and improve their social skills.* Champaign, Illinois: Research Press, 1975.

Liberman, R. P., King, L. W., & DeRisi, W. J. Behavior analysis and therapy in community mental health. In H. Leitenberg (Ed.), *Handbook of behavior modification and therapy,* Englewood Cliffs, New Jersey: Prentice-Hall, 1976.

Liberman, R. P., Fearn, C. H., DeRisi, W. J., Roberts, J., & Carmona, M. The credit-incentive system: Motivating the participation of patients in a day hospital. *British Journal of Social and Clinical Psychology,* 1977, *16*, 85–94.

Luber, R. F., & Hersen, M. A systematic behavioral approach to partial hospitalization programming: Implications and applications. *Corrective and Social Psychiatry,* 1976, *22,* 33–37.

MacDonald, K., Hedberg, A., & Campbell, L. M. *A behavioral revolution in community mental health.* Paper presented to the 5th Annual Meeting of the Association for the Advancement of Behavior Therapy, Washington, D.C., 1971.

MacDonald, M. L., & Tobias, L. L. Withdrawal causes relapse? Our response. *Psychological Bulletin,* 1976, *83,* 448–451.

Marmor, J. Comment. *American Journal of Psychiatry,* 1974, *131,* 780.

McFall, R. M., & Lillesand, D. B. Behavior rehearsal with modeling and coaching in assertion training. *Journal of Abnormal Psychology,* 1971, *77,* 313–323.

McFall, R. M., & Marston, A. R. An experimental investigation of behavior rehearsal in assertive training. *Journal of Abnormal Psychology,* 1970, *76,* 295–303.

McFall, R. M., & Twentyman, C. T. Four experiments on the relative contributions of rehearsal, modeling, and coaching to assertion training. *Journal of Abnormal Psychology,* 1973, *81,* 199–218.

Moos, R. H. *Evaluative treatment environments.* New York: Wiley, 1974.

National Institute of Mental Health. *Research in the service of mental health* (DHEW Publication No. ADM 75-236). Washington, D.C.: U.S. Government Printing Office, 1975.

Nietzel, M. T., Winett, R. A., MacDonald, M. L., & Davidson, W. S. *Behavioral approaches to community psychology.* New York: Pergamon Press, 1977.

O'Leary, K. D., & Wilson, G. T. *Behavior therapy: Application and outcome.* Englewood Cliffs, New Jersey: Prentice-Hall, 1975.

Ozarin, L. D., & Levinson, A. I. The future of the public mental hospital. *American Journal of Psychiatry,* 1969, *125,* 1647–1652.

Patterson, R. *Practical methods for establishing and maintaining token economies.* Springfield, Illinois: Charles C Thomas, 1976.

Paul, G. L. The chronic mental patient: Current status — future directions. *Psychological Bulletin,* 1969, *71,* 81–94.

Rachman, S., & Teasdale, J. *Aversion therapy and behavior disorders: An analysis.* Coral Gables, Florida: University of Miami Press, 1969.

Reiff, R. Social intervention and the problems of psychological analysis. *American Psychologist,* 1968, *23,* 524–531.

Rieder, R. O. Hospitals, patients, and politics. *Schizophrenia Bulletin,* 1974, Issue no. 11, 9–15.

Rimm, D. C., & Masters, J. C. *Behavior therapy: Techniques and empirical findings.* New York: Academic Press, 1974.

Schaefer, H. H., & Martin, P. L. *Behavioral therapy.* New York: McGraw-Hill, 1969.

Smith, M. B. The revolution in mental health care — a "bold new approach"? *Trans-Action,* 1968, *5,* 19–23.

Spence, J. T., Carson, R. C., & Thibaut, J. W. *Behavioral approaches to therapy.* Morristown, New Jersey: General Learning Press, 1976.

Spiegler, M. D. *School days — creditable treatment.* Unpublished manuscript, University of Texas, 1972.

Tobias, L. L., & MacDonald, M. L. Withdrawal of maintenance drugs with long term hospitalized mental patients: A critical review. *Psychological Bulletin, 1974, 81,* 107–125.

Turner, A. J. Behavioral community mental health centers: Development, management, and preliminary results. In R. A. Winett (chair), *Behavior modification in the community: Progress and problems.* Symposium presented at the 83rd convention of the American Psychological Association, Chicago, 1975.

Ullmann, L. P. *Institution and outcome: A comparative study of psychiatric hospitals.* New York: Pergamon Press, 1967.

Ullmann, L. P., & Krasner, L. *Case studies in behavior modification.* New York: Holt, Rinehart & Winston, 1965.

Wallace, C. J., Teigen, J. R., Liberman, R. P., & Baker, V. Destructive behavior treated by contingency contracts and assertive training: A case study. *Journal of Behavior Therapy and Experimental Psychiatry, 1973, 4,* 273–274.

Welch, M. W., & Gist, J. W. *The open token economy system.* Springfield, Illinois: Charles C Thomas, 1974.

Wolpe, J. *Psychotherapy by reciprocal inhibition.* Stanford: Stanford University Press, 1958.

Wolpe, J. *The practice of behavior therapy.* New York: Pergamon Press, 1973.

Zusman, J. Some explanations of the changing appearance of psychotic patients: Antecedents of the social breakdown syndrome complex. In E. M. Gruenberg (Ed.), *Evaluating the effectiveness of community mental programs.* New York: Milbank Memorial Fund, 1966.

3

Organization of the Therapeutic Milieu in the Partial Hospital

Stephen Washburn and Marguerite Conrad

Introduction

Any gathering of two or more individuals develops an interaction, a social atmosphere or pressure, with which new persons added to the system may unconsciously resonate or be consciously influenced. If the effect is to raise the individual's self-esteem, decrease symptoms, or increase social functioning, it is inferred that the context of the social system, the milieu, is therapeutic. If, on the other hand, the individual becomes more reclusive, persistently disorganized, or suicidal, the milieu can appropriately be labeled as antitherapeutic.

In describing the milieu of inpatient mental hospital units, Gunderson (1978) has identified five process variables that define the therapeutic functions provided by a milieu, namely: Containment, Support, Structure, Involvement, and Validation. In addition, we suggest that Negotiation is a sixth process variable that is unique to a partial hospital system.

The function of Containment is to preserve life, prevent destructive behaviors, and provide stability; Containment receives its maximum emphasis on an inpatient secure unit. Support refers to those conscious

Stephen Washburn • McLean Hospital, Belmont, Massachusetts 02178 and Harvard Medical School, Cambridge, Mass. 02138. **Marguerite Conrad** • Late of Boston University School of Nursing, Boston, Massachusetts 02115.

efforts by the social network to make patients feel comfortable and enhance their self-esteem; the function of Support is to make patients feel
less distress and anxiety. Structure includes all aspects of a milieu that
provide a predictable organization of time, place, and person, the function of which is to promote changes in symptoms and action patterns of
patients who are considered socially maladaptive. In considering these
three functions in a partial hospital program system, Containment is of
little significance except when provided as an outgrowth of the components of the milieu that relate to Structure and Support.

Involvement refers to those processes that cause patients to actively
attend to and interact with their social environment; the purpose of
Involvement is to utilize and strengthen a patient's ego and modify
aversive interpersonal patterns. Validation refers to a process in the
milieu that attends to individual treatment program planning and opportunities to explore areas of success or failure; the function of Validation is to affirm a patient's individuality.

These five functions are defined by Gunderson as necessary conditions for an in-depth milieu. Negotiation, a sixth variable which overlaps
with Involvement, is uniquely emphasized in the partial hospital system. The function of Negotiation is to provide the client with a partnership and interaction with his treaters whereby he formulates his needs
with the primary program planner, thus allowing for a legitimacy of the
client's desires and ownership of his care. This process is accomplished
through the initial diagnostic and assessment interview, the setting of
treatment goals by patient and treaters, and an evaluative process at
designated points in the treatment experience. This function under-
scores the need for the clinician to understand the patient's theory
concerning the nature of his illness and to accept the patient's individuality. The means of facilitating the negotiation process include involvement in decision making, ongoing consultation with the patient and
family, and clearly defined initial expectations and treatment goals.
Because a partial hospital system is midway in the spectrum between
inpatient physical structure and community freedom and complexity, it
may be able to offer a greater range of functions to affect the ego than
can the intrinsically more confining inpatient status.

This chapter will review types of social networks that can be said to
form therapeutic milieus. It will describe the social and ego capacities of
the individual that are especially affected by those unique social networks. Finally, the method by which a partial hospital staff may or-
ganize with its patients to make available such therapeutic functions,
cluster these functions for various diagnostic classes of patients and their
problems, and defines them through tracks or levels of treatment, and
sustains the delivery without staff "burn out" will be described.

Types of Milieu Organization

Therapeutic Societies

Over the centuries a number of efforts have been made by societies to care for their eccentric members by constructing a benevolent, comprehensive minisociety that offers protection, lessened expectations, and a chance for some degree of "healing" or natural restitution of symptoms. This minisociety also offers an opportunity for at least limited development, a degree of vocation, and a life-span. An outstanding example is the town of Geel in Belgium, where for centuries grossly ill persons have been given a place in normal households in order to take some part in the life of family and town. This community was initially financed by the church and later by the town; in modern times the Belgian government has accepted responsibility for its continuation. This comprehensive, low-expectation social structure was high in the delivery of Containment, Support, and Structure, all of which relieved symptoms and promoted growth for many people; for others, however, the opposite was probably true.

Healing communities have been described in religious contexts throughout the world. In biblical times, we note communities formed around Jesus as his charisma attracted followers who became active members of a circumscribed group. A paralytic at Capernaum, it is suggested, had developed a paralysis due to intrapsychic conflicts, not unlike a condition we might describe as hysterical paralysis. In this community at Capernaum the healing ascribed to Jesus suggests the necessity of a positive atmosphere to enhance cure; Jesus himself makes reference to "seeing their faith." Weatherhead (1951) suggests that the trustful expectancy of the followers greatly increased the healing potential. In this therapeutic milieu it can be suggested that Support, Validation, and Involvement were of primary importance in accomplishing the healing task.

In primitive cultures, figures such as the shaman play the important role of a healer who essentially provides a language by which the healing of psychic states in the community can be addressed. The shaman develops the individualized verbal or musical composition to assist the afflicted and members of the family through an episode of illness; this is a basic healing phenomenon. Help from other members of the society is of great importance to the shaman, who incorporates them into aspects of his healing rituals. This type of therapeutic community demonstrates a high level of Support and Structure with an added focus on Involvement.

Small circumscribed societies have been described by anthro-

pologists as practicing forms of magic and witchcraft when dealing
with the life crisis of individuals in a community. In these situations
difficult interpersonal relationships, despair, anger, and disease plunge
individuals and communities into crisis. There is a high value placed on
peaceful living; when this is disrupted, the sociotherapy of choice is the
practice of magic. The wizard or the witch in this healing community
becomes the primary practitioner and serves to provide the society at
large with ready-made rituals and beliefs to serve in critical situations. A
community that uses magic as a healing tool would tend to be especially
high on Support and Involvement, as is the case with religiously
oriented healing communities. They pay high regard to the input of
members of the afflicted network and describe the positive attitudes of
the persons surrounding the healer as enhancing and facilitating
growth.

For a decade or two following the institution of "moral therapy" in
America, mental hospitals provided a comprehensive agrarian work
program in benevolent, mildly expectant environments; this approach
seemed to promise remarkable success for psychotic patients. Unfortunately, this could not be sustained as hospital staffs became discouraged
with the unrelenting vulnerability of many patients. Imaginative Support and Structure gave way to a rigid, regressive Containment.

Modern-day attempts to recapture the creative aspects of moral
therapy and the atmosphere of Geel are seen in several New England
ventures. Spring Lake Ranch Therapeutic Community offers communal
living in a rural Vermont setting with the philosophy that shared work
builds relationships, prevents isolation, and is a vehicle for personal
growth. Ill persons are considered "guests," not patients, and are expected to become members of the "therapeutic community."

Farm Collaborative, located in a New England village, offers the
opportunity for family-style living in the environment of a working
farm. For the emotionally disabled individual, the farm milieu includes
training in the skills of daily living, a supportive atmosphere for personal development, and the opportunity to develop horticultural skills.
The farm staff participates side by side with members and real economic
proceeds from the work are divided among all participants.

The Task Oriented Community offers supervised residential living
for the ex-hospital chronic schizophrenic patient in an urban setting.
Along with the expectations of shared responsibility for cooking, cleaning, and house maintenance, a major focus is the reintegration of work
habits in the daily life of the residents. An adjacent assembly workshop
provides training and employment; residents are paid for each elec-
tronics component assembled. This sheltered residential work environ-

ment is unique in that it is not necessarily intended to serve as a transition to usual community living. Instead, the community can be a continuous "alternative" domicile for vulnerable people and thus may not stimulate the relapse rate (40%) engendered at Fairweather's Lodge (Fairweather, Sanders, Cressler, & Maynard, 1969). Both settings deliver the opportunity for involvement but the higher expectations of the Lodge for transition makes it, perhaps, less supportive of the individual's personality defenses.

In contrast to the supportive, healing, low-expectation models mentioned above, residential programs for drug addicts such as Synanon in California and Marathon House on the East Coast offer a highly organized milieu. Characterized by Bookbinder (1975) as "an intentional ethical community," each is staffed by drug-free ex-clients who use an intensely confrontive, educational approach that delivers Containment, Structure, some Involvement, and some Validation provided the individual complies with the Synanon-related work structure. Participants are encouraged to remain attached indefinitely and are subjected to heavy adverse pressure when they wish to depart. To obtain a balance between communal and societal demands, an educator was recently brought to Marathon House to assist participants in reorientation and transfer to the outer society. He assists members in meeting educational and vocational requirements for survival in the outside world.

The "therapeutic community" as it is popularly known today was the outgrowth of the concepts developed by Maxwell Jones. There is great difficulty in finding a rigid definition of a therapeutic community but according to Clark's (1965) critique, Jones's best definition is that "a Therapeutic Community is distinctive among other comparable treatment centers in the way the institution's total resources, both staff and patients, are self-consciously pooled in furthering treatment."

The therapeutic community was characterized by limited size, the regular occurrence of community meetings, and a basic underlying philosophy that is psychodynamically oriented. The therapeutic community as described by Jones uses many different methods for dealing with problems. Social methods that have had an outstanding role in the therapeutic community include freeing of communication, flattening of the authority pyramid, provision for learning experiences and role examination, and analysis of social events.

The most famous therapeutic community was founded in 1947 at the Henderson Hospital in Belmont, England. In this unit there were no manifest status distinctions (no uniforms were worn) and the clients and staff milled around talking freely in a fashion that made it impossible to distinguish roles. Clark (1965) summarizes Rapport's analysis of the

Belmont unit describing four themes that characterized the "unit ideology" as follows:

1. Democratization: Every person had one vote; everyone's opinion (nurse,
   doctor, or patient) was as good as the next
2. Permissiveness: The members were expected to tolerate disturbed behavior; discussion was better than discipline.
3. Communalization: Equality and sharing were valuable; everyone should
   express their thoughts and share them with others.
4. Reality Confrontation: All were expected to face their problems and in-
   terpretations were vigorously forced on them.

Social-Psychodynamic Organizations

Opportunities for Involvement and Validation are high in these
communities, whereas Support and Structure are given less emphasis.
The social system in which the patient is located contains a range of
staff–patient, patient–patient, and staff–staff interactions that offer continuing opportunities for some impact on the patients' ego, which can
favorably effect symptom remission and enhance personality growth.
Schizophrenic, borderline, and other patients with major ego defects
will recapitulate their pathological patterns with the staff in any setting.
Based on an understanding of the individual's psychodynamics, a staff's
stance may block or redirect the oral and aggressive drives of manic-
depressive psychosis, the needs for symbiosis and individuation in
schizophrenia, the externalization of self-directed aggression or the at-
tenuation of a primitive superego in depression, or repair a porous
superego in a character disorder. These maneuvers, described in more
detail elsewhere (Washburn, 1968) can augment or facilitate the effects
of intensive individual psychotherapy when used with these psychiatric
conditions. The goals and problems of psychoanalytical psychotherapy
in achieving improvement in the basic operations of the ego (from part
to whole object relations, self-object differentiation, lessening of the
defenses of denial and projection, a shift in ego identity, etc.) and the
way in which ego-defective patients can be assisted by therapist–staff
coordinations is aptly described by Vanderpol (1968).

Stanton and Schwartz (1954) demonstrate how misinformation,
disagreements with policies of the larger organization, and especially
covert disagreements between personnel tending the patient can contribute to a patient's pathological excitement or other psychotic states.
Edelson (1964) has theorized that a setting may achieve its therapeutic
effect on the individual's ego when it is altered by certain group characteristics such as goals, cohesiveness, standards, and redirection. Denber
(1960) has theorized that for a day hospital social milieu to have a beneficial effect, presumably in the area of self-image and interpersonal capac-

ity, the patient must experience a range of object relationships in a variety of groups. The effects of a large milieu therapy group in decreasing inappropriate instinctual expression on the treatment unit is described by Buck (1972).

The most purely ego-oriented focus is described by Cumming and Cumming (1967), who see ego growth occurring in a setting that exposes the patient to a spectrum of problems with which he can grapple. The staff both supports and role-models for the patient in his struggle to reality-test the ingredients of the problem and to develop new instrumental and social skills to cope with problems.

From this review, it can be seen that supportive-religious systems are high in the provision of Gunderson's functions of Containment, Support, and Structure. Jones's therapeutic community, Synanon, etc., additionally provide considerable Involvement. The psychodynamic organization emphasizes Involvement and individual Validation as well as an expectation of ego growth. These functions in the partial hospital milieu are further augmented by the demands of community living and the necessity to negotiate with staff, family, and vocational contacts. This expanded interface promotes the additional therapeutic function of Negotiation in the partial hospital milieu. To offer and maintain this range of functions an analogous Involvement, Validation, and opportunity for Negotiation is required of the staff itself. To demonstrate how this process can be initiated, two organizational models will be described and an example of a third model that has definite advantage in the milieu of a partial hospital system will be provided.

Staff Precursors of a Modern Milieu

In a milieu in which a range of services is offered the social system and the environment become the therapeutic agent. Milieus fall into a number of patterns from which variations occur within any given setting. The pattern that emerges tends to be determined by the framework of the service delivery system and the nature of the staff, particularly those in leadership positions. The milieu structure is determined by the mission or goals of the organization as established by those within a partial hospital system or by those who have an influential role in the parent or other impacting systems. Once the tasks and goals are defined there is an ordering of priorities, which then determines where the resources will be allocated as well as the method whereby the service will be carried out in order to achieve maximum performance and coordination. Finally, it is essential that policies and procedures be formulated in order to control both the performance and the coordination aspects. Whatever the philosophy of a milieu may be, there must also be a structure whereby functions are delineated, accountability is defined, and integration is facilitated. Specifically, within any partial hospital milieu it is important to be attentive to the goals of the organization, the tasks to be performed, and the processes whereby functions and accountabilities are integrated and defined. The psychiatric milieu is complex and its complexion is continuously changing. The arrival of a new patient, a group of training nurses, or a new psychiatric resident can change the community atmosphere in a partial hospital setting. In addition, a partial hospital system is not a contained milieu but rather one in which the organizational structure flows into the community at large. In that sense, it is inappropriate to think of milieu therapy in the same limited framework as is applicable to an inpatient system.

The "prescribed" milieu within a partial hospital system is one which tends to be mechanistic in quality and is most similar to the traditional medical model. Burns and Stalker (1961) describe the elements of a mechanistic system that are similar to those of a prescribed medical milieu. Within the medical model there is great emphasis on orderly lines of authority with importance placed on superior–subordinate interaction patterns. Usually, personnel tend to focus on the method of performing a task rather than on viewing the overall objective of the organization. There is a tendency to concentrate policy making in top administrative personnel with a restriction on the dissemination of information. Much of the communication in this system tends to be by the method of instructions. There is a preoccupation with the norms of loyalty and obedience as the means of maintaining positive membership within the organization. There is a tendency to focus on local standards of reference in evaluating the value of the members.

In the "prescribed" milieu there is also a great emphasis on the one-to-one relationship and programs tend to be highly individualized. Groups that develop for treatment purposes are highly structured, tend toward complex membership criteria, and have impermeable boundaries. The interactions between the milieu staff and the clients is ordered and lacking in spontaneity. The milieu staff will frequently seek decision-making sanctions from those in a leadership position and may in fact defer the whole decision-making process to those in the administrative posts. "What should I do about Mary Doe?" or "You'll have to ask your Doc" are frequent statements within this type of system. This milieu offers a high level of Containment, Support, and Structure.

The "pseudoegalitarian" concept translates itself into a milieu in which it is frequently said that roles are blurred and the general expectation is that anybody will do everything. The philosophy of many day hospitals (and much of the earlier interpretation of the therapeutic community) stressed the equality of roles and the desirability of all staff members working on an equal level. But in reality, equality does not exist and those who are less equal tend to identify the discrepancy. In the last analysis, job status and financial compensation are, in fact, unequal for different professionals and no instances of equality are described in the current literature. In the preoccupation with equality, the creativity of specialization has been lost.

A third model that best translates itself into the partial hospital system is the "egalitarian" model. In this model an open organic system as described by Burns and Stalker (1961) is appropriate. This model initially involves the identification of goal and purpose and an appreciation of task performance that respects the education, skills, and experience of staff members. It provides for decentralized decision making at the point of staff–patient encounter. Tasks are assigned according to skills and experience along with the responsibility of feedback to appropriate colleagues. There is definition of individual tasks as related to the whole system and the network of control; authority and communication is generated by individual motivation rather than on a hierarchical basis. There tends to be a high level of information sharing, diffusion of knowledge, and policy making with multiple lateral interaction patterns. Communication, a part of the feedback system, is high on information and suggestions and there is high value placed on quality of task performance and contribution to the goals and purposes of the organization. A cosmopolitan standard of expertise tends to emerge. Argyris (1970) suggests that such an open system is susceptible to change and members strive to accept every responsibility that helps increase their confidence in themselves, their group, and their capacity to solve problems effectively.

Developing an Open System

How then does one develop an open system that is highly productive and leads to a milieu that is therapeutic in nature? An example from the authors' experience will be used as an illustration of one possible methodology. Analogous to the system described by Jones (1976), who stresses the impact of organizational development techniques, we set about to redesign our own milieu. Four years ago our partial hospital system consisted of a day program that had been functioning for 10 years. It was decided to expand the program into a day, evening, night, and weekend program and concurrently develop a community aftercare component. The staff was grappling with the direction to take; there was disagreement, blaming of other persons, and a rather low morale factor particularly among the direct care givers, including mental health workers and nurses (i.e., the milieu workers). The mental health workers were caught in the familiar bind of being responsible for activity programs, motivating the patients, and generally keeping the center under control while the doctors and social workers were perceived as performing high-status work.

A decision was made to bring in a *change agent*, an outside consultant, to help evaluate the state of the organization. Initially we hoped to design and introduce a method of assessing the effect of the program on the partial hospital client. However, the consultant, after "scouting around" for 3 weeks, concluded that this was not our primary need. Rather, it was suggested that the focus of work become the development of shared priorities for patient services as well as the planning and implementation of psychosocial programs that reflected these priorities.

The change agent had to be constantly aware of the reactions of members of the organization, for according to Schein (1969), one of the most important sources of information about the readiness of an organization to change is found in the reception of and reaction to the consultant. The readiness of the staff to trust the consultant, their receptivity to new ideas, and willingness to share their concerns about the process of change are all important indicators. In addition, opportunities were created to identify the perceptions and desires of individual staff members regarding both their present and their ideal service priorities for the program. Finally, data were collected from patients about the strengths and weaknesses of the service they were currently receiving.

During the diagnostic phase a systematic process including questionnaires and interviews served to identify a number of concerns and issues for the staff. Staff tended to agree that they were at that time giving the most help to clients by prescribing medication, maintaining present levels of functioning, and helping clients talk about their feelings. However, when asked about how they perceived their ideal partial hospital center functioning they all agreed that medication would be less emphasized. While the staff saw the provision of a secure and comfortable place as an area of least help in their ideal center, they nevertheless agreed that at the present time the partial hospitalization program staff was giving the least help in the exploration of community resources as well as expectations of behavior change. In an ideal partial hospital center, the staff agreed that exploration of community supports would be the most helpful. Finally, there was staff disagreement as to whether low or high priority should be given to the involvement of the family, helping patients achieve insight, and helping patients talk about their feelings.

Of interest is the extent to which disagreements took place along discipline lines. It was especially notable that the psychiatrists rated insight and medication as very high priority while other disciplines focused more on psychosocial elements.

In the next phase, the staff collectively decided on a first priority from which emerged the design and planning of a comprehensive activity and rehabilitation program. Through this action, staff role ambiguities, conflicts regarding the status of activities and the talking therapies, and staff responsibilities were clarified. The objectives of the consultation project thus became (1) to increase staff awareness of the importance of addressing systems issues, (2) to change the range of priorities and expectations of the partial hospitalization service, (3) to implement decisions formulated on an executive level, and (4) to utilize staff time together at weekly meetings more effectively. The latter was of a highly significant nature in that it focused on the owning of feelings and opinions by all staff members and permitted the recognition and acceptance of conflict expression among staff members. Inherent in this activity was the encouragement of a norm of mutual staff support and the engaging of passive participants in the expression of feelings and opinions. Wise, Beckhard, Rubin, and Kyte (1974) have paid particular attention to the infrequent observation of these norms in health care system teams.

In summary, a group of staff members worked together toward the common goal of developing a therapeutic modality that would assist the client to achieve adequate and optimal functioning within the community at large. Can this goal be reached if the persons who make up the cadre of care givers are unable to define among themselves common values and goals? These goals and values of necessity must be real goals defined according to the needs of the client and incorporating input from this group; the goals may not necessarily be based on the values set by the most prestigious group or the most powerful persons within the system. In other words, milieu therapy involves ownership on the part of the client served, which is congruent with his/her goals. This ownership is unlikely to develop if staff members do not feel themselves to be partners in the development of the service to be provided. If a two-way communication process emerges that permits a satisfying interpersonal interaction between differing viewpoints, then mutual learning and support will be a component of that system. This is a learning environment that addresses challenges according to the "problem-solving method" as opposed to the "order and instruction" model.

At this point, it may be felt that a finished product has resulted and that the organic, egalitarian, organized milieu will be on its way to becoming and remaining a therapeutic tool. But is this really possible, given the attributes that have been described? In fact, there is no finality;

uncertainties continue, for goals and purposes are reordered and needs are endlessly changing. There is the continual wish for stability and reduction of uncertainty that, if it occurs, in all probability suggests that a closed system has evolved. Consider, for example, the community meeting that is a characteristic part of most milieu programs. Finally, the patient leadership is effective, programs are well under way, and there is a feeling of stability; at this point, the leadership group may suddenly begin the departure process through the attainment of jobs, community attachments, or a return to school. We find that within a short period of time there is, again, a community in search of leaders. The very healing process itself leads to a disruption in the community that serves a healing role.

Of importance in this discussion is the tension that may arise between the needs and goals of the various subgroups within a milieu. The client strives toward more independent functioning — to be healed. The staff has a vocational interest in which healing or service is a high priority. Job satisfaction and the pursuit of relevant interests must be given consideration. Recognition must be given to these concerns on all levels. As we noted earlier, within our own staff there were subgroups that identified different priorities and goals. Added to this system was a group of clients who come for service and who themselves had ideas about why they are there and what they might expect. The integration of these variables and the sense of negotiation within the subgroups and the larger client–staff group will be discussed below in more detail as we review how the functional effects of a milieu program are delivered to patient groups of differing ego capacities or pathological diagnoses.

Delivery of Therapeutic Functions to Patients

In preparing to make the previously defined functions a reality a staff should keep in mind certain basic issues, namely: (1) the type of patient, (2) the instinctual issues that are difficult for the patient at points in the treatment process, and (3) the types of social experiences (small groups, community milieu, task-oriented) that may be recommended to the patient. Since clients come to a partial hospital from either an inpatient unit or the community, the reactions to admission to a partial hospital program are varied. For the patient coming from the inpatient service there are termination issues with his former caretakers. Ambivalence about the need for continuing care is coupled with the temptation to flee completely. For the client coming from the community there are fears of emotional regression, and concerns regarding what will happen to him and whether acceptance by new caretakers will be forthcoming.

In our service an experienced psychiatric nurse functions as the receiver of each new patient and initially informs him about the program. Instead of asking something from the new patient, this individual gives something in the form of information and a concrete schedule of daily life, activities, and groups available within the center. Exploration of what the patient hopes to derive from the treatment sets the stage for the beginning of the *negotiation process*. The patient is introduced to other patient members and is encouraged to explore as a potential consumer the positive and negative aspects of the program. Following this the patient is asked to complete an assessment scale which assists staff and patients in defining the problems and the goals for treatment.[1] Administered prior to attachment, it serves to stimulate the client to think about those areas he wishes to work on as well as to identify skills and strengths. The patient identifies his goals and problems and has some framework within which these can be formulated. While staff may initially be fearful that patients will resist completing such an inventory, patients actually look forward to making this initial contribution to the planning and feel they are a potent part of that process.

Each client comes with a variety of needs, which are addressed through a mosaic of social, rehabilitative, recreational, and psychological approaches. Integrated within and derived from these approaches are the functional elements of Structure, Validation, Involvement, Support, and Negotiation described above. These are delivered through a level of care or *treatment track* appropriate for each individual. Stereotyping the client as acute or chronic might be misleading; considerations of ego strength and fragility, as well as social capacity, are more accurate.

At the point of entry into our partial hospital system, treatment tracks provide for differing levels of intensity determined by the effect of stress on the patient.

Maximum Track. For the client who enters the program at the point of experiencing an acute psychotic break or subacute decompensation, it is pertinent that emphasis be placed on a program that is high on Support and Structure and will help the patient feel comfortable and secure, resulting in a decrease of anxiety and stress. Use of executive ego functioning in the patient's area of strength and previous accomplishments is encouraged. Milieu approaches that provide service in this sphere are orientation meetings, attendance at structured community functions,

---

[1]The Maynard Personal Assessment Rating, an instrument designed and copyrighted by Lee Maynard, R.N., was implemented to assist the staff and patients to identify problems and goals for treatment at entry level and throughout the treatment experience.

and work and recreational programs. Clients in this track usually benefit from insight-oriented group and individual therapy, active vocational and rehabilitation counseling, and experiences within the community at large. The relatively short duration and high intensity of such a track is illustrated by the following example.

A 24-year-old woman was admitted to the partial hospital program complaining of anxiety, loss of weight, early morning awakening, and a suicidal gesture with aspirin 72 hours prior to admission. She broke up with her boyfriend of 3 years 1 month ago, and 1 week ago was fired from her job. She has had no previous psychiatric hospital experience but has been in individual therapy for 2 months. Because of suicidal potential her therapist felt she needed support beyond what is available in office practice. Miss J.'s goals on entry were symptom relief, regaining her former level of functioning, acquiring a job, and gaining understanding of her interpersonal difficulties, especially with men.

Initially, Miss J. met with an admitting nurse, who informed her about the program and assessed her present degree of stress through an assessment instrument evaluating mental status. Based on the patient's goals and this initial evaluation, the orientation group and community activities (including a work program in the community) were recommended. The orientation group provided a support group with other newly admitted patients and a teaching mechanism whereby Miss J. could be helped to understand and make full use of the system. The work program consisted of a group that twice weekly performs tasks for the handicapped and elderly in the community. Participation in this group helped Miss J. regain self-esteem through work that was of practical value and from which she received immediate feedback. At the unit community meeting and coffee hours she joined with other members to plan activities and discuss daily life issues. Attendance at a team meeting three times weekly provided an arena for exploring the interpsychic issues leading up to her current crisis and negotiating for and developing her plans for regained independence. This dovetailed with a heightened focus in her continuing individual therapy. This maximum program, high in Support and Involvement, continued for 3 weeks on a 5-day-per-week basis. As her depression lessened she acquired a part-time job 3 mornings per week and in 5 weeks was coming to partial hospital program 2 days per week. As she became more independent, Miss J. increased her time at work and by the 8th week was involved in group therapy on a once-weekly basis. Through this program, which focused on verbalization of problems, strengthening of executive ego functioning, and recognition of interpersonal relationship difficulties, the patient returned to full functioning with the support of an ongoing group and outpatient treatment with her individual therapist. High emphasis was placed on helping the patient feel competent, respectable, and potent in the environment (i.e., validated).

Maximum Maintenance Track. This low-expectation track, high in frequency of attendance, addresses a group of patients who have a long-standing psychiatric disability, lack a social network outside the hospital, are dependent on the institution, and have no marketable skills. This patient would be a typical long-term inpatient if a partial hospital were not available. Clients in this group are exceedingly high on resistance to involvement, which is notable in their lack of interaction with their social environment. Since many clients in this group operate on the fringe of their capacity, individualized, tailored programming permits exploration of areas of possible competence. This group of patients has experienced failure on many levels and opportunity for resolving symptomatic reactions to these events is of prime importance. As the patient's individuality is affirmed, there is greater capacity for closeness and self-understanding. The following example will illustrate the use of the maximum maintenance track with this patient type.

A 35-year-old male was admitted to the partial hospital program following a short-term inpatient experience during which he expressed paranoid ideation and stayed on the fringe of the inpatient milieu. On initial contact the nurse found him to be guarded, uncertain of his goals, and hesitant to become involved in any activities. Although he was unable to verbalize his goals, the client's responses on the assessment instrument indicated that he wanted to learn more about the management of his own finances and wished to have an opportunity to be with people on a recreational level. He joined a community group that planned and participated in activities in the community. Part of his membership requirements consisted of sponsoring and planning one group activity per month. Through this activity he experienced the validation process that served to affirm his individuality and motivated him to operate in a minimally independent fashion. A skills of daily living group provided him with an opportunity to learn and practice skills such as food shopping, banking, and transportation opportunities; it was both supportive and involving in function. The client participated in this activity until he had mastered the skills on a satisfactory level. At the end of his experience, he was a coleader of the activity with the staff. A prevocational group addressed the job-related issues and he began to attend a supervised job corps. As he experienced ambivalence and uncertainty related to independence in the vocational area, he entered this phase of his program on three different occasions; it was 9 months later that he was able to participate in a limited workshop experience. Negotiation and reordering of goals were of particular importance in this case, for ambivalence related to work had to be addressed. The patient, in interaction with the primary treaters, defined his individuality and accepted the responsibility for and the legitimacy of his desires. Membership in the partial hospital community was enhanced through participation in a day community meeting; the patient initially participated in those aspects of the program that related to activities; however, he stayed on the fringe in verbal activities. A community activity recreational program, coupled with skills of daily living, a prevocational group, and flexible involvement in community activities on a long-term basis resulted in the patient gradually building a social network and participating in a sheltered workshop situation. His attendance in a dynamic program aided his slow progression toward independence.

Minimal Maintenance Track. This is designed for clients who possess some vocational skills, have a minimal social network, and experience independent functioning on a limited basis. These patients usually come to the program from a more intensive treatment level with a set of specific goals. There is a high priority placed on ownership and formulation of needs as well as the need to evaluate progress at vital points in the program. Through involvement the patient moves to modify aversive patterns and strengthen his/her functional ability. Experiences that contribute to this ideal are assertiveness training groups, interpersonal involvement, and self-awareness; this approach is enhanced by participation in individual programming. Negotiation at this point affirms the integrity of the individual through attention to the ambivalence, resistance, and tolerance for human expression of feeling as the improved patient disengages from the partial hospital program. The following example will illustrate this description.

A 26-year-old woman had entered the partial hospital program 6 months previously following a manic-depressive illness. Again gainfully employed as a receptionist, a job that was satisfying to her, she lacked any social network outside her family constellation. In reviewing her current treatment goals she explored the need to develop some meaningful social relationships but was unable to proceed beyond the verbal level. An assertiveness training group allowed her an opportunity to develop some beginning social skills in a peer group. In addition, she attended a weekend program that focused on planning and implementing an activity with members of the program. After 4 weeks she was ready to invite a friend from work to dinner; she successfully completed this project. Gradually she was able to translate social skills learned in assertiveness training and planning and practice groups to her own daily living. At this point, she was able to join recreational and socialization groups in the community and come to the center once a month for monitoring of medication.

Minimum Track. This program also addresses the needs of clients who are stressed periodically by the emergence of acute problems; these problems interfere frequently but minimally with functioning and during their occurrence require Structure, Support, Negotiation, and Validation. Those patients who are unable to make a more in-depth commitment fall in this category. They usually have adequate vocational skills, function relatively independently, and have a social network. Their need for service is usually crisis-oriented.

Another use of the minimal track is by a group of seriously impaired persons who on entry can tolerate no initial assignment to groups. They are unable to relinquish control or trust members of a treating team and yet are capable of some alliance. When an attempt is made to treat these patients the Negotiation process becomes the crucial element of the treatment program; it is here the flexibility of the team approach is most severely taxed. The following example will illustrate the use of the minimum track.

An 18-year-old male who suffered psychotic symptoms since childhood was admitted to a specialized day school of emotionally disturbed adolescents. A backup psychiatric service was required to provide structure and support as a supplement to the school experience.

On entry his anxiety and fear prevented him from attending any groups within the milieu. However, with the individual nurse practitioner he was able to state that he was lonely and wanted to learn how to get along with other people and understand why he was having difficulty with interpersonal relationships.

A nurse orchestrated his program and designed it to include limited contact with a variety of professionals who were maximally supportive. Additionally, students were helpful in reinforcing this staff response utilizing behavior modification principles. The social worker and nurse meet with the client's parents to extend the milieu program to the community and provide consistency with the partial hospital focus. Six months after admission the client was able to tolerate attendance at team meetings and gradually participated in the socialization group. In the latter group he pondered his inability to make friends; in the current events group he practiced the art of conversation. Bouts of disruptive, overbearing behavior were rare and were limited by sending him home rather than to an inpatient unit. The milieu thus provided a degree of containment.

Major Modalities That Make Up the Track System

As was noted earlier, each client comes with a category of needs that are addressed through recreational, social, rehabilitational, and psychological approaches. The programs are oriented toward tracks not unlike those described in the educational models and entry level prescription is in accordance with the functional abilities of each individual.

Recreational needs of a diversified client population are addressed by developing a spectrum of activities (Figure 1) that lead to value definition and skill building. For the long-term client the Outbound I program provides an opportunity to participate in activities in the community that may range from bowling to boat trips. The beginning phase consists of a planning meeting in which patients are expected to make a contract or commitment of up to 6 weeks and agree to plan and sponsor one activity. To be a sponsor entails working cooperatively with fellow group members in exploring common interests and making arrangements for an activity. Included in the group are staff members who provide guidance and support and facilitate implementation of ideas. The activity itself is supplemented by discussion in which members are encouraged to explore expectations and interpersonal issues related to the actual activity. It is expected that members will then take this experience and begin to independently plan recreational opportunities in their community of origin.

For a younger population more intensive activity programs in the community are provided and, again, an emphasis is placed on interpersonal relationships that evolve from these activities; in addition, an opportunity is given to analyze behaviors that interfere with the successful attainment of recreational skills. For many partial hospital program clients the weekend provides a major source of stress resulting from the inability to develop structure related to recreational needs and a lack of individually defined interests. For this reason a group that focuses on weekend planning and provides an opportunity for feedback and assistance in pursuing recreational opportunities is provided. For those unable to put these ideals into practice, a Saturday afternoon planning and activity group helps the individual develop and practice these skills with a group of people in the treatment network.

A transitional group in the recreational sphere is the leisure-bound group, which actually marks the final step between the activities in the partial hospital program community and a growing dependence on the community at large. The client in this program is afforded an opportunity to assess leisure values. The client experiences the removal of the therapist as facilitator and is expected to develop leisure activities on his own initiative. The client assesses what he/she likes to do and when to do these activities. These recreationally oriented programs provide the client with an opportunity to develop executive ego skills through opportunities for leadership. Planning within the groups is focused on the interpersonal relationships and each individual is encouraged toward independent functioning according to his capacity.

For the partial hospital client who has a deficit in *socialization*, skills groups are developed that provide for individual self-awareness, the development of interpersonal skills, and the organization of social skills.

The orientation program for the new client coming into the system focuses on concrete information sharing about the program as a whole. The client is encouraged to explore issues related to transitioning into the program and fears related to being a patient, and is helped to understand the ground rules of the center. There is no time limit or contractual expectation and in some instances it is indicated that clients who have been withdrawn and fearful stay with the group for an extended period of time. For the client who lacks confidence it may be indicated that he/she become the coleader and therefore develop a responsibility in reaching out to new patients coming into the system.

A socialization group focuses on issues related to social interactions and assumes that clients will make an 8-week contract. Clients are given homework assignments; videotaping is utilized to give immediate feedback to clients in the group setting.

Other groups with more specific goals and purposes are included in this area. Progressive relaxation focuses on self-awareness and self-control of tension and anxiety through the utilization of systematic exercises. Creative movement focuses not only on self-awareness through individual exercise but also on interpersonal awareness through group exercises that accommodate both verbal and nonverbal skills. An assertiveness training program is designed to facilitate cognitive awareness of interpersonal transactions through the use of specific skills that are then applied to increasing interpersonal effectiveness. Homework assignments and a contract are expected for membership in this group. A current events group provides opportunity to develop the art of conversation around topics of unusual interest; the goal is to provide clients with an opportunity to think together about issues that are related to ongoing events in the world at large and to become competent in the ability to converse with one another.

To function as a valued member of Western society, each individual is expected to maintain a minimal level of productivity and failure to do so implies a great stigma. There is a lack of acceptance implied in the inability to work and be productive. A *rehabilitation* spectrum provides opportunities to engage in a work experience on a variety of levels based on the functional abilities of the client. For the low-expectation group, participation in a recycling program on a twice-weekly basis provides the opportunity to develop minimal work skills and an ongoing experience as a member of a work team. In time, a patient, with support of the

Monday	Tuesday	Wednesday
Team Meeting 9:00–10:00 M.D., R.N. M.S.W., M.H.W.	Outbound II 8:00–1:00 M.H.W.	Socialization 9:00–10:00 R.N., R. T.
Outbound I 11:00–4:00 M.H.W., R.T.	Team Meeting—Team I 9:00–10:00 M.D., R.N., M.S.W., M.H.W.	Coffee ½ Hour 9:00–9:30 M.H.W.
Orientation 11:00–12:00 R.N.	Recycling 10:00–11:30 Volunteers	Play Reading 10:00–11:00 Patient Volunteers
Community Meeting 12:00–12:50 R.N., M.H.W.	Green House 10:00–11:30 Volunteers	Opera Appreciation 11:00–12:00 Music Therapist
K.P. 1:00 & 4:00 Team II	Community Action 12:30–3:00 Sr. M.H.W.	Newsletter Committee 12:00–1:00 M.H.W.
Art Group 1:00–2:30 M.H.W., Students	Recreational Activity Group 2:00–3:00 R.T., R.N., & Students	Current Events 12:20–1:00 M.D., Sr. M.H.W.
Creative Movement 2:30–3:30 M.H.W.		K.P. 1:00 & 4:00 Team I
		Recreational Activity Group 1:30–2:30 Staff & Students
		Team Meeting 3:00–4:00 M.D., R.N., M.S.W., M.H.W.
		Leisure Bound 4:00–5:00 R.T., Rehab.
A.I. Group 6:15–7:15 M.D., R.N.	Outbound 6:00–9:30 M.H.W., Volunteers	Assertiveness Training 6:00–7:00 R.N.
Women's Group 8:00–9:00 R.N.	Craft Workshop 6:00–8:30 O.T.	Art Studio 7:00–9:00 A.T.
	Progressive Relaxation 7:30–8:30 M.D.	Job Group II 7:30–8:30 Voc. Rehab.

Code to Abbreviations Used:
    M.D. — Psychiatrist   M.S.W. — Social Worker
    R.N. — Nurse   M.H.W. — Mental Health Worker

Figure 1. Therapeutic activities schedule.

Thursday	Friday	Saturday
Breakfast Group 8 30–10 30 Sr M H W	Team Meeting 9 00–10 00 M D , R N , M S W M H W	Special Program A I Group 9 00–11 00 M D , R N
Green House 10 00–11 30 M H W	Job Group I 10 30–11 30 Voc Rehab	Group Therapy 12 00–1 00 M D , R N
Poetry Reading 10 00–11 00 Patient Volunteers	Photography 10 30–12 00 M H W	Outbound Planning 1 00–1 30 R N , M H W
Community Action 12 30–3 00 Sr M H W	Community Meeting 12 00–12 30 R N M H W	Outbound Activity 1 30–4 00 R N , M H W
Craft Workshop 1 00–2 30 O T	Recreational Activity Group 1 00–2 00 Student Nurses	
Weekend Planning 2 00–3 00 M H W	Skills of Daily Living 1 00–3 00 Sr M H W	
Recycling 3 00–4 30 Volunteers	Diet Group 2 00–2 30 M H W & Dietician Consultant	
Medication Teaching R N , M D	Craft Workshop 3 00–4 00 O T	

A I Group 6 00–7 00 M D , R N	Assertiveness Training 6 00–7 00 R N	Dinner Group 5 00–6 00 R N , M H W
Craft Workshop 6 00–8 30 O T		
Outbound 7 00–9 30 M H W		

R T — Recreational Therapist   A T — Activities Therapist
O T — Occupational Therapist

Figure 1 *(Continued)*

staff, may become the crew leader and facilitate involvement in the project. It is helpful if this leadership position interdigitates with a special interest on the patient leader's part.

An assignment to the work crew on the grounds under extensive supervision provides experiences that are varied in nature and are of minimal demand on the patient as he moves to acquire basic work skills and habits. During this assignment he may be involved in a painting project, cleaning a wall, or projects of a similar nature.

A community action group assists the patient in his work-related activities to move into the community and begin to develop a limited role of responsibility. Activities that are generally pursued during this stage are voluntary in nature and may include working with older citizens in the community, perhaps doing some painting, cleaning a yard or raking leaves, or reading to a blind student. Of great importance is the fact that those in the community who experience this service tend to be very grateful and feedback is highly gratifying in nature. In our experience patients have been highly motivated by this mechanism. The work projects are negotiated by a staff member who must serve as a facilitator throughout the project. A side effect of this experience has been an increasingly positive reaction to our patients as a group and to mental patients in general.

More challenging experiences are available in structured work situations, which include food service in a coffee shop and attendance in community-based workshops. Involvement in these programs is coupled with intensive rehabilitation counseling either on a one-to-one basis or in a group. As a final move the patient may assume a job in the community and attend a job discussion group that addresses issues related to ongoing problems in the work situation.

Finally, there are those programs within the milieu that serve to hold the partial hospital community together. Community meetings on a regular basis are necessary and become the arena whereby staff and clients work out "town meeting" issues. Input is garnered from the various team meetings, the tracks, and the individual patient. In the community meeting anyone may bring up a personal or group problem and it is expected that the management of problems that arise in the center over an extended period of time will also be addressed. Since passes, privileges, and the like are not relevant in a partial hospital setting, issues arise that relate to occurrences or difficulties in interpersonal relationships, developments in various programs, and issues related to the work of the treatment teams. Patients routinely relate to themes developing in the team and compare how these themes may be relevant to the community at large.

The Central Role of Negotiation in the Partial Hospital Milieu

"The ego has incorporated a particular bit of behavior when a person can forget what he is doing and instead find out what he can do with it" (Erickson, 1950). In reviewing theoretical considerations in the development of a milieu, Erickson's statement sums up the goals of a program designed to address the needs and goals of patients or clients and stress the highly relevant process functions of Validation, Involvement, and Negotiation. Learning takes place through various ranges and depths of intimate personal relationships.

We are often faced with the unmotivated patient who will not follow his program or take advantage of what is offered within the milieu. And yet this individual represents the stark need for ego integration and growth in the healing community. To deal with these issues within a milieu we must be mindful of the process Orlando (1961) calls the "situational conflict." In discussing nursing activities the author points out that when the nurse, in collaboration with the patient, ascertained what would meet the patient's needs and acted upon this knowledge, the patient's behavior or condition improved. In contrast, "it was noted that activities decided upon without adequate exploration were inconsistent with the patients' needs." To expect the activity to benefit the patient without negotiation is not realistic. When solutions evolve from the unresolved "situational conflict" activities are ineffective in bringing about the desired results and there is an impression that the patient is unmotivated, not behaving in a healthy way, or being a troublesome client. Experience within our own milieu confirms that if program direction or activity is not redirected or addressed through the process of reflection and renegotiation of ineffective activities, further delays in improvement result. We again refer to that process in the milieu whereby the function of negotiation is essential if the patient's goals for treatment are to be met. When the issue of control and power within a milieu supplants the process of negotiation we have a setting ripe for the situational conflict; requirement and justification become the predominant responses because maintaining one's position as client or staff becomes the primary concern directing our efforts. On the contrary, our expertise can be translated into a situation in which the negotiation process creates an atmosphere of trust and fosters the emergence of mutuality. Must there be blatant change or can we be satisfied with subtle changes? When change is demanded, the client feels threatened, defensive, and rushed and is without the freedom and time to affirm and understand new learning: "Pressure to change without an opportunity for exploration and choice seldom results in experiences of joy

and excitement in learning" (Morimoto, 1973). Finally, we must re-
member that people are filled with anguish in the process of change; we
must be able to understand and accompany them in these periods of
great pain. In other words, as milieu workers we must affirm the integration of negotiation and empathy.

References

Argyris, C. *Intervention theory and models*. Reading, Massachusetts: Addison-Wesley, 1970.
Bookbinder, S. M. Educational goals and schooling in a therapeutic community. *Harvard Educational Review*, 1975, 45, 1.
Buck, R. E. A large milieu therapy group. *American Journal of Psychotherapy*, 1972, 26, 3.
Burns, T., & Stalker, G. M. *The management of innovation*. London: Tavistock, 1961.
Clark, D. H. The therapeutic community-concept, practice and future. *British Journal of Psychiatry*, 1965, 3, 947–954.
Cumming, J., & Cumming, E. *Ego and milieu: Theory and practice of environment therapy*. New York: Atherton Press, 1967.
Denber, H. C. B. *Therapeutic community research conference*. Springfield, Illinois: Charles C Thomas, 1960.
Edelson, M. *Ego psychology, group dynamics, and the therapeutic community*. New York: Grune & Stratton, 1964.
Erickson, E. H. *Childhood and society*. New York: W. W. Norton, 1950.
Fairweather, E. D., Sanders, D. H., Cressler, D. L. & Maynard, H. *Community life for the mentally ill*. Chicago: Aldine, 1969.
Gunderson, J. G. Defining the therapeutic process in psychiatric milieus. *Psychiatry: Journal for the Study of Interpersonal Process*, 1978, 41, 327–335.
Jones, M. *Maturation of the therapeutic community: An organic approach to health and mental health*. New York: Human Sciences Press, 1976.
Morimoto, K. Y. O. Notes on the context of learning. *Harvard Educational Review*, 1973, 43, 2.
Orlando, I. J. *The dynamic nurse–patient relationship: Function, process and principles*. New York: G. P. Putnam's Sons, 1961.
Schein, E. H. *Process consultation: Its role in organization development*. Reading, Massachusetts: Addison-Wesley, 1969.
Stanton, A. H., & Schwartz, M. S. *The mental hospital*. New York: Basic Books 1954.
Vanderpol, M. The designed milieu as an extension of the psychotherapeutic process. In S. H. Eldred & M. Vanderpol (Eds.), *Psychotherapy in the designed therapeutic milieu*. Boston: Little, Brown, 1968.
Washburn, S. L. Milieu interventions in the treatment of psychosis. In S. H. Eldred & M. Vanderpol (Eds.), *Psychotherapy in the designed therapeutic milieu*. Boston: Little, Brown, 1968.
Weatherhead, L. D. *Psychology, religion and healing*. Nashville, Tennessee: Abington Press, 1951.
Wise, H., Beckhard, R., Rubin, I., & Kyte, A. L. *Making health teams work*. Cambridge: Ballinger, 1974.

II

Special Populations

Introduction

The numerical growth in partial hospitalization has been accompanied by a concurrent expansion in the types of patients treated within the general context of part-time programs. Although some success has been achieved by the incorporation of patients into heterogeneous programs, it is generally being recognized that patient types with particular characteristics and problems may require unique approaches or, at the very least, special consideration within existing programs. It is becoming increasingly important, therefore, for treatment agents to be cognizant of the special considerations involved in dealing with unique patient populations.

In this section, the special treatment needs and conceptual considerations involved in dealing with two specific populations are described. First, Benjamin Lahey and David Kupfer detail the treatment of children and adolescents in the context of partial hospitalization; advantages of part-time institutional care and specialized treatment techniques required in light of the characteristics of this population are discussed. Second, John Neisworth and V. DeCarolis Feeg consider the unique conceptual and clinical problems involved in the treatment of the mentally retarded; the authors place partial hospitalization within a continuum of treatment possibilities and provide guidelines for programming.

4

Partial Hospitalization Programs for Children and Adolescents

Benjamin B. Lahey and David L. Kupfer

Introduction

The confluence of several important contemporary trends has recently given strong support to the development of partial hospitalization programs for children and adolescents. Partial hospitalization programs provide institutional care and treatment for exceptional children on a part-time basis (less than 24 hours per day), while the child remains in the community the rest of the time. This treatment strategy has received strong support from three movements within the human services field: (1) Under the banner of "deinstitutionalization," nearly every state has mandated that as many full-time residents of institutions as possible be returned to the community; (2) an increasing body of clinical research strongly suggests that parents can be trained to be effective therapists even with very deviant children, and that successful parent-therapists apparently can have even more durable and pronounced effects on the maladaptive behavior of their children than professional therapists who work with the child outside of the family setting (O'Dell, 1974); and (3) there is increasing realization of the need for families to have regular relief from the care of exceptional children. This need to be relieved from full-time parenting is particularly great if there are other children in the family, if both parents are employed, or if the exceptional child belongs

Benjamin B. Lahey and David L. Kupfer • Department of Psychology, University of Georgia, Athens, Georgia 30601 to a single-parent family (one including a single, widowed, or divorced parent). In one way or another, all three of these trends suggest that partial hospitalization programs will play an increasingly important role in the care and treatment of exceptional children and adolescents.

The term *partial hospitalization* is actually a misnomer. It incorrectly implies that only hospitals use such programs. In actuality, a wide variety of facilities conduct part-time treatment for children and adolescents, including community mental health centers, university psychology clinics, child guidance clinics, day care centers, halfway houses, and even special schools for exceptional children. It is the federally funded community mental health centers that have spearheaded the recent spread of these programs, in an effort to move away from what Silver (1977) calls the "all-or-none" model toward the more comprehensive "continuum of care" model offering outpatient, part-time, and inpatient services for young people. Perhaps more importantly, the term *partial hospitalization* gives the impression that only children who are suffering from "diseased" minds are served by the programs, with all the problems attendant to the medical model of maladaptive behavior. It would probably be better to replace "partial hospitalization" with a more neutral term such as *part-time institutional treatment*.

Types of Programs

A large assortment of programs fit within the general definition of part-time treatment. The most common are programs that treat the child during some part of the daytime hours (mornings, after school, or all day) 5 or 6 days per week. Such programs are usually referred to as day treatment programs or day hospitals. Other models of part-time treatment include afternoon and overnight care for youths who are attending public schools or holding jobs but who cannot live with their families. Certain programs provide only emergency or "respite" 24-hour residential care in the event of family crises or extreme difficulties with the child's behavior. In one sense, each of these models provides a different kind of service, but they all share the common feature of an institution sharing the responsibility for the care and treatment of an exceptional child with the family and outside community. They also share the goal of returning the children they treat to their families and communities as soon as possible.

The day treatment program described by Marshall and Stewart (1969) is typical of those programs that provide services on a weekday basis. Some children and adolescents come to the treatment center from 8:30 A.M. to 5:00 P.M. Monday through Friday. They attend a special school at the center during the mornings and participate in recreational activities, discussion groups, and therapy in the afternoon. Others attend public schools in the community and come to the center only for the afternoon activities. Parents are involved in the program by being required to provide overnight care, clothing, transportation, and other custodial services. They receive regular family therapy, which is believed to indirectly benefit the child, but they are not trained to be active therapists with their children.

Harris (1974) has developed a day treatment program at the Rutgers University Child Behavioral Research and Learning Center that exemplifies programs for very deviant children. Ten children, aged 5 through 12 and diagnosed as autistic, are served 5 days per week. The core of the program is an educational program based on behavior modification procedures. All educational and therapeutic efforts are geared toward teaching the children whatever skills they will need to return to a special education class in their own school district.

In contrast to the program discussed by Marshall and Stewart (1969), the Rutgers program utilizes the parents as therapeutic resources. Parents are required to attend weekly behavior modification workshops, read parent-training guidebooks, and are trained to design practical treatment procedures to deal with their child's behavior problems in the home. Still, treatment was primarily carried out in the day school by the professional therapists and aides, who in this case were specially trained college students. An evaluation of the treatment outcome of the program was encouraging, but data were not provided on comparable children who were treated in other ways.

A very different model of part-time institutional treatment was designed by Drabman and his associates (Drabman, Spitalnik, Hagaman, & Van Witsen, 1973). Referred to as the "five-two" plan, the program provides 24-hour residential care and treatment 2 days per week (usually on the weekend), while the child remains in the home full time for the rest of the week. The plan is designed to provide relief and assistance to the *family*, which, in contrast to other programs, is seen as the primary source of treatment and care.

The five-two split of residential responsibility is actually the final phase of an integrated strategy that is designed to train the parents to be behavior therapists. Initially, the child is kept in the residential center full time for 3 to 4 weeks. During this time, the child's behavior is assessed, intensive treatment is begun in the center by the professional staff, and most importantly, the parents are given *intensive* training in behavioral treatment methods in the residential center. The facilities of the treatment center are used to allow the parents to actually practice the treatment methods under the supervision of the professional staff. This allows the staff to train the parents to a high level of competence in a relatively short period of time.

After 3 to 4 weeks the child is returned to the parents' care 2 days per week. Gradually, the amount of time spent in the home is expanded as the parents become more skilled as therapists for their child. In some cases, the child also comes to the center to attend a special day school during the weekdays. For significantly deviant exceptional children (such as those labeled moderately and severely retarded or autistic), this plan seems to provide an ideal sharing of the responsibilities for treatment and care between the institution and the family.

Along similar lines, Astrachan (1975) has proposed a model in which the child or adolescent spends 5 days per week in full-time residential care but stays with the family on the weekends. Her program involves the parents in treatment to a lesser degree but incorporates the same concept of shared responsibility between family and institution.

While some day treatment centers offer children a rather loosely scheduled therapeutic milieu, the program administered by the Behavior Research Institute, in Providence, Rhode Island, serves as an example of a more structured approach. As described by Freeman, Ornitz, and Tanguay (1977), this program ambitiously attempts to serve a population between the ages of 9 and 21 who enter treatment with a wide range of diagnostic labels. These young people attend a day school 6 full days a week, 12 months a year. The school operates from a behavioral orientation, featuring carefully individualized treatment packages for each student and his or her family. Students can progress at their own educational pace through personalized systems of instruction, using self-paced programmed textbooks and teaching machines. Conduct problems are often handled by negotiating contingency contracts between the school and the student, making specified rewards dependent upon appropriate behavior. A token economy system is used to motivate students to master the skills they personally need in order to return to the public education system. Daily behavior charts keep track of each student's progress through this program.

Proposed Advantages of Part-Time Institutional Care

Many writers have presented arguments in favor of the use of part-time institutional treatment for children and adolescents. Although different models of part-time treatment appear to offer different advantages, the supporting arguments can be summarized as follows:

1. *Part-time institutional treatment is more economical than full-time residential care.* Because most models of part-time treatment do not re-

quire the existence of true residential facilities (dormitories, bathing facilities, etc.), because all part-time models require the employment of fewer professional staff, and because the families are usually required to bear most of the cost of food, clothes, and transportation, part-time treatment programs require the expenditure of fewer tax-generated funds. This may mean that limited financial resources can be used to support other programs that would not otherwise receive support, and inpatient beds can be freed for disturbed youngsters who truly need 24-hour care. Administrators, moreover, have found that a part-time program can be added on to an existing residential program with minimal additional cost, since many of the physical facilities and staff members can be shared.

This proposed advantage has been placed first in the list, however, because it requires our greatest scrutiny. Whenever an alternative treatment strategy is proposed that costs significantly less than traditional methods, there is the probability that it will receive the strong endorsement of administrators and political leaders. This means that perhaps the strongest impetus for the newly proposed treatment strategy will be largely independent of the effectiveness of the program. It is possible under such circumstances that a new program that was initiated for its therapeutic promise will be continued for political and economic reasons.

An analogous situation can be found in public school programs for exceptional children. When the mainstreaming concept was proposed, the goal was to provide more normal social learning experiences for exceptional children. In this model, the children stay in regular classes as much as possible, attending resource classes when they need special instruction. This model has been subverted in many school systems, however. Instead of hiring the number of special education teachers needed to adequately serve the exceptional children in a system, some administrators have used the mainstreaming model to provide only token service to these children. In some cases, children who need several hours of special instruction per day are receiving 1 hour per week. This allows administrators to hire one-fourth of the number of special teachers that could be effectively used. In the same way, precautions must be taken to see to it that inexpensive part-time treatment programs are not used to inappropriately substitute for other necessary, but more expensive services.

2. *Part-time programs provide needed relief to families.* Every advocate of part-time treatment has emphasized the benefits of the program to the clients' parents and siblings. In most cases, the parents of significantly deviant children simply need a rest from the 24-hour-a-day, 7-days-a-week responsibility for their children. Felner, Stolberg,

and Cowen (1975) have shown that children raised by separated, divorced, or widowed parents are especially likely to become maladjusted. With a part-time institution temporarily sharing their parenting duties, these single-parent families can more effectively adjust to their difficult situations. While their deviant child is being aided by part-time treatment, parents may find the time to improve their relationship with other children. It may also give the parents enough distance from their child to clarify their feelings and their commitment to his or her care. All of these gains for the family, of course, could indirectly benefit the client, but that should not be a necessary requirement for providing part-time treatment. There is a growing trend to treat all members of a family as clients, regardless of which member was initially identified as the one with psychological problems. By this logic, in a situation in which part-time treatment would neither help nor harm the original client, part-time treatment could be justified if it would benefit the other members of the family. Still, the primary purpose for removing the child or adolescent from the family on a part-time basis is to benefit the target client.

3. *In contrast to full-time institutionalization, part-time treatment allows the family to remain intact.* Placement in a full-time residential facility necessitates almost total separation from the family. In most cases, this deprives the child of a major source of support, positive role models, and enjoyment. Furthermore, this type of abrupt separation may function as a source of stress at a particularly vulnerable point in his or her life. Moreover, as Freeman (1959) has suggested, the families of children who are given full-time residential treatment may "reorganize" in their absence. These possible changes in relationships and routines might make it difficult for the child to return to the family after discharge from residential treatment. Part-time treatment programs, in contrast, allow the child to remain in the intact family on a regular basis. In this manner, the existing strengths present in the family can be made part of a child's treatment plan. In addition, the part-time treatment model enables a child to learn new social skills during the day and try them out with his family that evening.

4. *Part-time treatment programs can partially remove the child from maladaptive family environments.* In some cases, the value of part-time treatment derives primarily from the time spent out of the family, rather than the time spent in the family. In cases where it appears that the disruption, inappropriate models, and counterproductive reinforcement contingencies of the family are largely responsible for a child's problems, full-time institutionalization should be considered. But even when a family seems extremely chaotic, there may be reasons to avoid communicating such a total and permanent loss of faith to the child's natural

family. In such cases, part-time treatment provides a new environment in which more appropriate forms of behavior can be learned.

5. *The part-time treatment model provides an excellent opportunity for families to learn new ways of acting toward their exceptional child.* While all proponents of this model stress the importance of parental involvement, the nature of this involvement varies greatly. Some programs offer traditional modes of family therapy, in which professionals help the family members understand and alter problematic family relationships. Many part-time programs involve groups of parents meeting together regularly, discussing common experiences and problems. Other programs take a more concrete parent-training approach, training parents in the use of management techniques proven effective in dealing with disturbed children. Staff members can model appropriate ways of acting with a child, and then teach the parents to apply these skills when the child is at home. When this type of parent training is successful, the child who shows improvement at the part-time center will also show improvement at home.

6. *Part-time treatment will not necessarily disrupt a child's education.* Prolonged residential placement, in automatically removing a child from his classroom, may be subtracting a source of stability as well as interfering with intellectual growth. A part-time program is capable of offering a child therapeutic services while simultaneously allowing him to continue in a familiar classroom in his own community.

7. *Part-time treatment will not necessarily disrupt a young person's social life.* This point is especially valid for adolescents, passing through a stage in which belonging to a group may be all-important. The stigma of full-time residential placement may make adolescents feel like social outcasts. Part-time treatment, however, can allow them a degree of social freedom balanced by a certain amount of institutional protection. The part-time program staff can teach the youths interpersonal skills and then encourage them to use these skills to relate better with peers in their neighborhoods.

8. *Part-time treatment programs can be used to avoid some of the iatrogenic effects of institutionalization.* Although full-time placement in a well-designed and -executed residential treatment facility is necessary and even advantageous to the client under some circumstances, full-time placement carries with it the danger of iatrogenic, or treatment-induced, harmful effects to the client. The sterile, monotonous environment of the institution, the primacy of the need to maintain institutional order over the need to provide treatment, the shortage of professional staff, and the side effects of medication and other forms of institutional therapy are a few of the well-known sources of probable iatrogenic

effects. Even if the child or adolescent improves enough to be returned to the community, the mere fact that the individual has been away for a stay in a "mental hospital" may make it difficult for him or her to be accepted again by teachers, peers, and even family members. The availability of a part-time program may serve to prevent, replace, or shorten a child's stay in a long-term institution.

9. *Part-time treatment is preferable to parents of exceptional children.* Not only does part-time treatment avoid the negative effects of institutionalization on the child, but it is often a more acceptable alternative to parents. Parents often experience considerable guilt over full-time residential placement, so much so that it can interfere with the return of the child to the family. Parents who feel that they "caused" their child to be institutionalized, for example, may become overly indulgent and permissive with their child after he or she returns from full-time institutionalization. Part-time programs can be especially effective in lessening parental guilt by actively involving parents in their treatment plans.

10. *Part-time programs can be used to avoid the stress of separation of child and family.* Much has been learned about the effects of stress on children over the past 30 years, especially from the work of Bowlby (1969). One thing is clear: When children are under severe stress, their maladaptive reactions will be exacerbated if they are separated from their families. In recent years, Fraser (1973) has documented this phenomenon by describing the increase in psychophysiological and anxiety reactions among children who were removed from battle zones in Belfast, Ireland, to safe areas away from their parents.

Since most institutionalizations occur during times of crisis, it seems reasonable to assume that some children and adolescents will be cut off from their primary source of support when they need it most. While the family itself is the "battleground" in many cases, even here, total and sudden separation from the family may not be desirable.

11. *Part-time programs can be used to ease the transition between outpatient and full institutionalization.* Evangelakis (1974) has suggested that part-time programs can be used in a number of ways to facilitate the transitions between outpatient and residential care for those individuals who require some period of full-time residential treatment. The most common transitional use of part-time treatment is between full-time and outpatient care. Many such programs are currently in effect with adolescents, most often referred to as "halfway houses." They are designed to ease the client back into the family and community, and in many cases serve treatment functions as well. Residential treatment programs that are successful in helping children and adolescents behave in prosocial patterns cannot automatically be counted on to result in a maintenance

of the new behavior patterns after discharge. On the contrary, it seems necessary that youths will have to be taught adaptive ways of living in the community just as they were taught to behave adaptively in the hospital. A halfway house can teach a client's family and friends how to react to the client in ways that will maintain the new gains in appropriate behavior.

Atthowe (1974) has designed a transitional day treatment program for psychotic adults based on this premise. Clients who have been released from full-time residential care are taught to find employment and form appropriate social relationships while still in a partial institutionalization program. Later, after family members and employers have been trained to maintain such gains, the program aids in the transition to full-time community living. Similar programs can, of course, be developed for children and adolescents.

Evangelakis (1974) has also suggested that part-time treatment programs can be used to ease the transition from family living to full-time residential treatment. He suggests that part-time treatment should be tried first in an effort to avoid full-time placement entirely. Even in cases where full-time residential care seems inevitable, an initial period of day or part-week treatment could be used to ease the transition to residential care.

12. *Part-time institutional programs can provide the opportunity for extended assessment.* If assessment is conceived as acquiring in-depth information about an individual's typical patterns of behavior and the situations in which that behavior occurs, part-time institutional programs provide an opportunity for extended close observation under carefully controlled stimulus situations.

Part-time institutional programs provide both the advantage and disadvantage of being able to observe the child outside of the normal family and school environment. As Marshall and Stewart (1969) have suggested, having the child out of the home for prolonged periods of time allows the child to be "untangled" from the possibly maladaptive relationships of the family. On the other hand, this type of assessment *cannot* supplant assessment of the child in the family and school environments that may have caused or at least maintain the child's maladaptive behavior. Assessment must include both behavior of the client and the social and physical environments in which the child behaves. Assessment in the new environment of the day treatment center of the child's maladaptive behavior provides exceptionally useful assessment information, but will almost certainly lead to an inaccurate assessment if it is the *only* source of information. Still, the information gathered in several weeks of systematic observation is immeasurably superior to that gathered in a 50-minute interview or a battery of written tests. The day program designed by Thompson, Garrett, Striffler, Rutins, Palmer, and Held (1976), in fact, accepts some preschool children solely for prolonged assessment 3 mornings a week for 6 to 8 weeks.

Although it is not commonly done, part-time institutionalization also offers the opportunity to assess the impact of the client on siblings and parent. To some extent this information can be gathered retrospectively, but perhaps the best approach would be to assess the behavior of all members of the family both before and after part-time treatment has begun.

13. *Part-time programs can be used as the base for treatment.* The primary reason for the existence of part-time programs is, of course, to provide treatment. The sheer number of hours that the child or adolescent spends with the professional staff in contrast to outpatient treatment programs provides the opportunity for both intensive and extensive treatment programs. Depending on the orientation of the staff and the nature of the children, these could include a variety of individual, group, and educational treatments.

But, if treatment is confined to the child during the time he or she is in the treatment center, the long-term effectiveness of the program will almost certainly be negligible. As mentioned before, adaptive changes in behavior cannot be expected to generalize to situations outside of the treatment center unless steps have been taken to ensure that they will do so. This is simply because individuals behave in ways that are appropriate to the situation. If they are encouraged and supported for behaving in appropriate ways in the treatment center, they will be likely to do so, but if maladaptive behavior is encouraged in the family, that will be the behavior that predominates. O'Leary and Wilson (1975) have suggested that self-reinforcing behavior changes, such as relief from anxiety and some newly acquired social skills, can be expected to generalize to nontreatment settings, but patterns of aggression, crime, substance abuse, and psychotic behavior probably cannot be permanently changed without intervention outside of the treatment center.

For this reason, the part-time center must be conceived of as the *home base* of treatment interventions, rather than as the sole location for treatment. This means that the treatment staff must serve as consultants to parents, teachers, parole officers, and others as much as or more than they function as deliverers of treatment directly to children. As the medical model of abnormal behavior has declined in the human services profession, we have become less confident that treatment can "cure" the mind of deviant individuals. In that outdated conception, it is easy to believe that the individual would take his or her newly healthy mind, and hence appropriate behavior, into any new situation. It now seems apparent that to achieve improved behavior in the home, school, and other major environments, treatment must be carried out at least on a consultative basis in each of those environments.

The experience of Lovaas and his associates (Lovaas, Koegel, Simmons, & Long, 1973) in treating psychotic children speaks directly to this issue. Initially, his highly successful behavior modification program was carried out in a full-time residential setting. This was apparently chosen because it allowed the therapists to gain the maximum degree of control over the therapeutic environment. After significant gains had been made in modifying the children's psychotic behaviors, they were discharged either to other full-time state institutions that employed traditional psychiatric methods or to their parents, who had been trained in the use of behavioral methods.

Long-term follow-up evaluations of these children revealed marked differences in the durability of the improvements in behavior. The children who had been discharged to state hospitals uniformly relapsed, frequently to levels of psychotic behavior that were higher than before treatment. In contrast, the children discharged to their families generally maintained their gains *or even improved*, particularly those who lived with parents who had sucessfully mastered the use of behavioral techniques.

These results led Lovaas to significantly restructure the nature of his treatment program. Although the behavior modification methods employed in the program were essentially unchanged, Lovaas switched from a full-time residential model to a day treatment program. In this program the child was brought to the center for treatment during the day but primarily resided with the family. The emphasis was not on receiving professional treatment in the day center, however, but on providing an opportunity to train parents to modify their own children's behavior. Since only those children whose parents had developed behavioral parenting skills had continued to improve, the new emphasis is on the intensive training of parents as therapists.

The children in Lovaas's program spend so little time in day treatment, in fact, that it could alternatively be considered an extensive outpatient program. The children are brought to the treatment facility only as frequently as is necessary for the training of the parents; in fact, as progress is made with the parents and child, the frequency of visits is reduced. The best features of this program could be combined with the best features of other part-time treatment programs. For example, the children could attend the day program on a regular 5-days-per-week basis or spend 2 nights per week in the center on a 24-hour basis, to provide assistance to families in the care of their deviant children.

Education in Part-Time Treatment

Educational programming constitutes a particularly important issue when considering part-time treatment for children and adolescents. The children referred to day treatment are often those who have had problems in public school during the day. Since these school difficulties often involve behavioral factors, part-time facilities must offer school services that are therapeutic as well as educational. In some models of day treatment, all children involved in the program attend a school that is under the direction of the treatment facility. In other programs, certain students attend classes in the public school system.

Providing schooling for children and adolescents in a day treatment center is a formidable challenge when one considers the characteristics of these young people. Instruction must be geared toward a wide variety of intelligence levels. While a large proportion of the children can be expected to be of below-normal intelligence, some may be exceptionally bright. Many children will have already experienced the special education labeling process, carrying designations such as learning-disabled, mildly retarded, brain-damaged, and behavior-disordered. A history of truancy is especially common for adolescents referred to day treatment. For all these reasons, children and adolescents come to day treatment from a background of school failures. The staff's task is to provide them with successful school experiences, preparing them academically and behaviorally to return to regular classrooms.

Much emphasis must be placed on helping children develop the preacademic skills that are prerequisite for success in school. Special attention is given to very basic areas such as self-help skills, language development, and speech therapy. If a child can master these fundamentals while in a day treatment center, he or she will be better equipped to learn actual academic material in school. The day treatment center classroom is also an excellent place for children and adolescents to learn social skills. A lack of social skills may have been a major factor in their previous school failures. Teachers can take part in helping them find better ways of playing and talking with peers. In addition, Koret (1973) has observed that many children in part-time treatment units appear to have difficulty relating to authority figures, and can specifically benefit from help in learning to take directions from teachers.

In order to reach these educational and psychological goals, class size in the part-time institution is kept as small as possible. Evangelakis (1974) discusses grouping children into small classroom groups not just according to their academic achievement level but also by their social-behavioral compatability. The classroom teacher must be part of a treatment team, working with other staff members to thoughtfully coordinate curriculum planning with the school officials who will be teaching educational and therapeutic goals. To ease the transition from a treatment facility back to the community, the teacher must also coordinate curriculum planning with the school officials who will be teaching the child once he returns to his own community.

Appropriate Clients for Part-Time Treatment

Although much has been written about the advantages of part-time institutional treatment, very little attention has been paid to the question of which types of clients would be best served by this type of program. Presumably the choice among full-time, part-time, or outpatient treatment should be made on the merits of each individual case, but can an argument be made that certain types of clients are most likely to benefit from part-time treatment? Based on a review of the characteristics of children served by the many part-time programs described in the literature, one reasonable answer would seem to be no. Virtually every diagnostic category of deviant children and adolescents has been served in one center or another. This is not to say, however, that each center serves a wide variety of individuals. It appears that numerous programs were designed to serve only a specific population (those who have normal IQs, who are diagnosed as "psychotic," "mentally retarded," etc.), while ignoring the other deviant children in the community. Many day treatment centers prefer to admit only those children who seem capable of working smoothly in groups with other children.

It appears, therefore, that virtually any category of exceptional children can be served in part-time treatment programs, provided the program has been designed to flexibly meet their needs. But after we go beyond such general groupings, are there specific client variables that predict success or failure in part-time treatment programs? Unfortunately, there is little information upon which to answer this question.

Freeman (1959) has stated that the children best suited for part-time treatment are those who seem to have high potential for psychological, social, and intellectual growth, but are currently functioning at a severely disturbed level. He has mentioned childhood schizophrenics as an example of a diagnostic group that could especially benefit from part-time treatment.

Several other authors have mentioned specific types of young people likely to gain from part-time programs. Ahmed and Stein (1974) have designed a part-time treatment package for adolescents who are on court probation. Referred from the courts, these adolescents are required to spend a portion of each day, after school, at a treatment center. They attend group therapy sessions and receive social skills training, aimed at preparing them to relate to others in socially appropriate, non-criminal ways. Braga and Braga (1975) have discussed the use of part-time centers to coordinate services to pregnant teen-agers. Both before and after childbirth, these women could receive medical, psychological, social, and continuing educational help through a part-time institution. Adolescents who have run away from home, or who have a drug abuse problem, have traditionally received medical, psychological, and social assistance at centers that fall within the general definition of part-time treatment.

Marshall and Stewart (1969) have proposed a set of guidelines for the selection of clients for part-time programs. An informal evaluation is made of the "strength" of the child, the family, and the community (teachers, principals, probation officers, peers, etc.). If the combined strength of these factors is high, outpatient treatment is recommended; if the combined strength is moderate, part-time treatment is recommended; and if the combined strength is low, full-time residential treatment is recommended.

While the concept of "strength" is unnecessarily vague, it seems clear that the family and significant members of the community must be able to provide an environment that encourages and supports adaptive behavior, either currently or after training, if part-time or outpatient treatment is to have a chance of being successful in the long run. It can be argued, however, that the strength of the child is not always a significant factor. Considerable research has demonstrated that when the strength of family and community resources are high, even children diagnosed as psychotic or severely retarded can be successfully treated on a part-time or even an outpatient basis. Clearly, however, some characteristics of the child or adolescent, such as serious aggression that the parents cannot control, must be taken into account when choosing treatment strategies.

What is most needed at this point is a clearer specification of what is meant by the "strength" of child, family, and community resources. Lovaas et al. (1973), for example, have suggested that the parents who are most likely to serve as effective therapists for their own children have three characteristics: (1) They believe that their children are not "sick" in a permanent sense, and that treatment will lead to significant improvement; (2) they are willing to use firm methods of discipline, such as time-out, loss of rewards, and even punishment when appropriate; and (3) they are willing to devote a major portion of their lives to the care of the child.

Much more discussion and research are needed on these issues, particularly on how client-related variables interface with different patterns of strength. For example, it would appear that a day treatment or five-two program would be very useful for parents who are not willing or able to devote their entire day or week to their exceptional children, and might also be used to demonstrate to skeptical parents the improvements that are possible through proper treatment.

Empirical Basis for Part-Time Treatment

While the variety of arguments that have been put forth in support of part-time treatment seem both sound and compelling, it must not be forgotten that they are no more than logical arguments. Before we can confidently recommend part-time programs, they must also be empirically supported. Have part-time programs received such support? In this case the answer is an emphatic no. Although some indirect evidence can be cited to support certain elements of the logical arguments (e.g., Fraser's data on the effects of separating children and parents during high-stress periods), there have apparently been no direct experimental tests of the comparative effectiveness of treatment programs. Such a test might consist, for example, of a study in which children with approximately the same type of severity of behavior problems are randomly assigned to either full-time, part-time, or outpatient treatment groups. If meaningful measures of maladaptive behavior are taken before and after treatment, comparison of treatment outcome could be made with at least some degree of validity. While there are many ethical and conceptual problems with such a study, the difficulties inherent in making treatment decisions in the absence of empirical evidence seem far more salient.

Other Alternatives to Full-Time Residential Treatment

In the search to find an alternative to full-time residential treatment, we must be careful to avoid unthinkingly choosing between full- and part time treatment. Rather, we should constantly reevaluate all available alternatives to see if other programs might more effectively meet the needs of our clients. In other words, we must be able to *flexibly* develop ways of sharing the responsibility for treatment among the family, the community, and the institution.

A model treatment program that illustrates this point is the Achievement Place concept for delinquent adolescents developed by Wolf's research group (Phillips, Phillips, Fixsen, & Wolf, 1971). They developed a model that combines institutional, community, and home resources in a way that differs significantly from any of the previously mentioned programs. Initially, the youth is assigned to essentially full-time residential treatment. The nature of the residential treatment is markedly different from most existing programs, however. Seven or eight delinquent boys or girls live in each family-style Achievement Place home. A married couple who are professionally trained "teaching parents" live with youths on a full-time basis and administer the treatment program. The youths attend school in the community and can go on earned trips home, to town, or to special entertainment events, but they otherwise spend all of their time in the home at the start of the program.

Each Achievement Place home is governed by a token economy that explicitly specifies the rewards for adaptive behavior and the penalties for maladaptive behavior. Contact is maintained with the school through direct consultation and specially designed daily report cards that earn, or lose, points in the home's token reinforcement system. Parent involvement begins early in the program and takes the form of training them to eventually take over treatment when the youth returns home.

In the beginning, the youth is on a strict reinforcement system in which earned points are traded for reinforcers (such as recreation, privileges, snacks, allowance, savings bonds, and trips away from the home) on a daily basis. As soon as the individual's behavior has improved enough to warrant it, he or she graduates to a status in which privileges can be earned or lost for a full week at a time. After 4 consecutive successful weeks on the weekly system, the youth advances to a merit system in which all privileges are free. If, however, there is a return to inappropriate behavior at any point, the youth is returned to the weekly or daily system until they can earn their way off again.

After 4 successful weeks on the merit system, the youth advances to the "homeward bound" system. During this period, the youth is gradually faded back into the family, who ideally have begun to acquire the skills necessary to maintain the gains made at the Achievement Place home through parent counseling. With the help of the teaching parents, the family sets up rules of behavior, privileges, and penalties similar to that at the home. The youth spends 1 or 2 24-hour days per week with the family in the beginning, and if successful, gradually moves to full-time family living. In the same way, the amount of support and training provided to the family by the home staff is intense at the start and gradually tapers off to nothing over a period of about a year.

Evaluations of the effectiveness of this program have been highly favorable. In comparison to a similar group of delinquents who were placed in traditional full-time institutions, or who were placed on court probation, the graduates of the Achievement Place homes have been apprehended for committing crimes less than half as frequently as institutionalized youths and almost one-third as much as youths placed on probation. In addition, the small family-style Achievement Place homes are less than half as expensive per capita as large state institutions.

The Achievement Place model differs significantly from most day treatment programs yet shares the common feature of shared treatment responsibility. Although it currently serves only delinquent youths, it may also be possible to treat other categories of exceptional children, especially if the program is modified to include some of the features of other part-time treatment programs. Again, however, our decision should not be to adopt a given model of treatment to serve a population, but to flexibly combine the elements in seemingly the best way possible, and then to experimentally evaluate our programs. The Achievement Place program, then, represents a model effort to flexibly develop effective treatment methods that are scrupulously subjected to empirical validation.

Summary

Part-time treatment programs appear to offer an economical way of providing improved treatment to many deviant children and adolescents. Such programs promise to ease or avoid the stress of family separation, support family efforts, avoid the iatrogenic effects of residential care, and facilitate broad-spectrum assessment and treatment. They strive to accomplish this by providing an optimum combination of the efforts of the family, the community, and the institution. The weight of the arguments in favor of part-time treatment is considerable, but there is virtually no empirical basis upon which to support its use. Clearly research is urgently needed on this topic.

We must be careful, in addition, not to allow arguments in support of any particular model of treatment to supplant our efforts to individualize treatment methods for our circumstances and clients. Perhaps the most important cautionary note that can be added to any discussion of treatment methods, however, is a call for greater commitment to nontreatment psychological services. We must not allow any treatment program, no matter how promising, to interfere with our community-wide efforts to *prevent* psychological problems.

References

Ahmed, M. B., Stein, D. D. Children's mental health services: A case study of a successful grant proposal. *Hospital and Community Psychiatry*, 1974, 25, 591–595.

Astrachan, B. M. The five-day week: An alternate model in residential treatment centers. *Child Welfare*, 1975, 54, 21–26.

Atthowe, J. M. Behavior innovation and persistence. *American Psychologist*, 1973, *28*, 34–41.

Bowlby, J. M. *Attachment and loss*. New York: Basic Books, 1969.

Braga, L.D., & Braga, J. L. Child development and community mental health services: An important partnership. *Child Psychiatry and Human Development*, 1975, *6*, 47–54.

Drabman, R., Spitalnik, R., Hagaman, M. B., Van Witsen, B. The five-two program: An integrated approach to treating severely disturbed children. *Hospital and Community Psychiatry*, 1973, *24*, 33–36.

Evangelakis, M. G. *A manual for residential and day treatment of children*. Springfield, Illinois: Charles C Thomas, 1974.

Felner, R. D., Stolberg, A., & Cowen, E. L. Crisis events and school mental health referral patterns of young children. *Journal of Consulting and Clinical Psychology*, 1975, *43*, 305–310.

Fraser, M. *Children in conflict*. London: Secker and Warburg, 1973.

Freeman, A. M. Day hospitals for severely disturbed schizophrenic children. *American Journal of Psychiatry*, 1959, *115*, 893–898.

Freeman, B. I., Ornitz, E. M., & Tanguay, P. E. Educational approaches at the Behavior Research Institute, Providence, Rhode Island. In E. R. Ritvo (Ed.), *Autism: Diagnosis, current research, and management*. New York: Halsted Press, 1977.

Harris, S. L. Involving college students and parents in a child-care setting: A day school for the child with autistic behaviors. *Child Care Quarterly*, 1974, *3*, 188–194.

Koret, S. The children's community mental health center emerges. *Child Psychiatry and Human Development*, 1973, *3*, 243–254.

Lovaas, O. I., Koegel, R., Simmons, J. Q., & Long, J. S. Some generalization and follow-up measures on autistic children in behavior therapy. *Journal of Applied Behavior Analysis*, 1973, *6*, 131–166.

Marshall, K. A., & Stewart, M. F. Day treatment as a complementary adjunct to residential treatment. *Child Welfare*, 1969, *48*, 40–44.

O'Dell, S. Training parents in behavior modification: A review. *Psychological Bulletin*, 1974, *81*, 418–433.

O'Leary, K. D., & Wilson, G. T. *Behavior therapy: Application and outcome*. Englewood Cliffs, New Jersey: Prentice-Hall, 1975.

Phillips, E. L., Phillips, E. A., Fixsen, D. L., & Wolf, M. M. Achievement Place: Modification of the behaviors of predelinquent boys within a token economy. *Journal of Applied Behavior Analysis*, 1971, *4*, 45–59.

Silver, L. Difficulties in integrating child and adolescent training and service in a community mental health center. *Hospital and Community Psychiatry*, 1977, *28*, 26–30.

Thompson, R. J., Garret, D. J., Striffler, N., Rutins, I. A., Palmer, S., & Held, C. S. A model interdisciplinary diagnostic and treatment nursery. *Child Psychiatry and Human Development*, 1976, *6*, 224–232.

5

Partial Hospitalization for Mentally Retarded Citizens

John T. Neisworth and V. DeCarolis Feeg

Introduction

Partial hospitalization in the psychiatric and mental health context has been recognized as a plausible alternative to services facilitating an individual's reentry to his home, community, or work after a period of hospitalization. Day hospitals and night hospitals have emerged along the continuum of rehabilitative services offering to appropriate patients a prescriptive and regulated blend of medical and therapeutic care with normal life (Astrachan, Flynn, Harvey, & Geller, 1970). These programs are not without disadvantages, although the advantageous outcomes appear to predominate (Guy, Gross, Hogarty, & Dennis, 1969). Some arguments arise criticizing the unnecessary duplication of halfway houses. The support groups, however, contend that the partial hospitalization concept can provide short-term intensive treatment during acute stress, where the halfway house is more suited to recovered patients who need a supportive setting over a period of time (Beigel & Feder, 1976b).

Like their nonretarded peers, retarded citizens sometimes require hospital care. When this is necessary, several issues arise related to the selection among options for such hospitalization. These major issues and their implications are discussed below.

John T. Neisworth • Division of Special Education, The Pennsylvania State University, University Park, Pennsylvania 16801. **V. DeCarolis Feeg** • Division of Biological Health, The Pennsylvania State University, University Park, Pennsylvania, 16801.

Background Considerations in Hospital Programming for Retarded Citizens

Health versus Intellectual Status

First, the frequency, duration, and kind of hospitalization that may be necessary for a retarded individual varies more as a function of individual circumstances than of "being retarded." Persons who are considered retarded — by whatever criteria — are hospitalized for reasons of *health* and not because of a condition of retardation *per se*. Once a person is engaged in a hospital program because of a health problem, there are, however, considerations in the care and treatment of retarded as opposed to nonretarded persons. These considerations will be discussed later in the section dealing with guidelines for hospital programming.

Health Related to Intellectual Status

It is true that the degree of health impairment is loosely related to the level of retardation. So, for example, mildly retarded individuals evidence essentially no more health problems than do nonretarded peers from similar (usually low) socioeconomic circumstances. Among the majority of our retarded citizens, then, hospitalization is no more necessary than among the nonhandicapped population. Nearly 80% of those who are considered mildly retarded by conventional psychometric standards evidence no unusual organic dysfunction, nor do they suffer significantly more chronic or sporadic health maintenance problems. The severely and profoundly retarded, however, do suffer a much higher incidence of acute and chronic organic involvement and related health dysfunctions. Usually, moderate and severe retardation is associated with medical syndromes that include health maintenance problems. It is for these persons of moderate to severe retardation that hospitalization is relevant and necessary. It happens, then, that the greater the retardation, the greater the need for medical and, possibly, hosptial care.

Biology versus Behavior

A third consideration associated with hospitalization for retarded persons is the "biological versus behavioral development noncongruence" principle (Neisworth, 1977). Briefly, this refers to the fact that while biological development and increasing behavioral sophistication are *related*, they are not inextricably tied to each other. We see in non-retarded children rapid progress in biological development and corresponding increasing behavioral development. But these two aspects of development are not so interdependent that the one cannot necessarily proceed without the other. That is, an individual may evidence great delays in behavioral development while at the same time experiencing normal biological maturation. The case is most clear when we consider the relatively normal timing for the onset of puberty among many retarded children and, at the same time, their lack of social behavioral sophistication to match their new biological attributes. Thus we must not *forsake* or neglect developmental progress in behavior because of retardation in biological status. If biological/behavioral status are seen as separate, albeit related dimensions, then intervention in the one need not necessarily impede progress in the other. Accordingly, treatment aimed at ameliorating biological or health problems must be designed so that it will not interfere with progress in learning. *Medical and hospital care designed to remediate health problems must not interfere with ongoing behavioral development.* At best, medical intervention can actually collaterally assist behavioral elaboration while rectifying health dysfunction. Or, medical care may be neutral with respect to behavior development; it may neither help nor hinder. At worst, medical intervention can ostensibly improve health status while actually deteriorating overall development. This iatrogenic effect has been noted by others (Stuart, 1970) in reference to the boomerang effect of treatment.

Influence of Contemporary Social Philosophy

Fourth, a contemporary view and philosophy of retardation provides implications for the choice and design of hospital programs. These implications are included as part of a discussion of two dominant themes in the treatment of retardation. These two rather basic and sweeping movements, brought about by a number of social forces, are normalization and its educational correlary, mainstreaming (Nirje, 1969; Wolfensberger, 1972).

Normalization

As both a process and a goal, normalization dictates that retarded persons must be progressively moved to more normal levels of functioning through successively more normal means of treatment. Thus, retarded citizens are expected to assume more community involvement, be seen as citizens of the community, and be integrated with nonretarded peers in most activities. The relevance of the principle of normalization to hospital care will be discussed later.

Mainstreaming

Implementation of normalization principles within an educational context is termed *mainstreaming*. Currently, handicapped children are rapidly being enrolled in school along with nonhandicapped children in common settings. Special schools, special classes, and other particularized segregated arrangements are being supplanted by integrated accommodations. By legal mandate, retarded children *must* be provided with public schooling *regardless* of the level of retardation or health impairment. Accordingly, new roles for teachers, counselors, nurses, parent educators, etc., are emerging to meet the tremendous demands that arise as a result of the new mix of students in the schools.

Interactional View

Implicit in normalization as a goal is the conviction that retardation is dynamic, changeable, and open to remediation — a distinct departure from the past. There is, then, an optimistic interactional developmental view of retardation (e.g., Bijou, 1977) that has great bearing on the issue of hospitalization.

Whether retardation is believed to have arisen as a result of constitutional or environmental factors or as an interaction of these, a prime consideration in the psychosocial treatment of retardation is the modification of the environment that the person routinely encounters. Numerous projects — both pioneering ones and current research — demonstrate the positive effects of changes in the quality of environment. Since approaches to both the origin and the treatment of retardation include consideration of the environment of the person, the quality of the hospital context becomes important to consider. That is, a hospital environment may promote or impede development, depending on characteristics of the program and setting.

Conventional and Emerging Hospitalization Options

The concept of partial hospitalization for the mentally retarded lends itself to a wide array of interpretation. The field of hospital care for all groups of individuals has been in active debate and remains unresolved. The introduction of the normalization principle has a revolutionary effect on the orientation of therapies for the mentally retarded. Habilitation for the mentally disabled through hospital care cannot be considered normal in any context.

Figure 1 is a time-framed description and hierarchical arrangement of the array of direct services for the mentally retarded. Its purpose is to represent the concept of partial hospitalization along the service continuum from most restrictive to least restrictive. A few definitions may help clarify various aspects of this perspective of therapies.

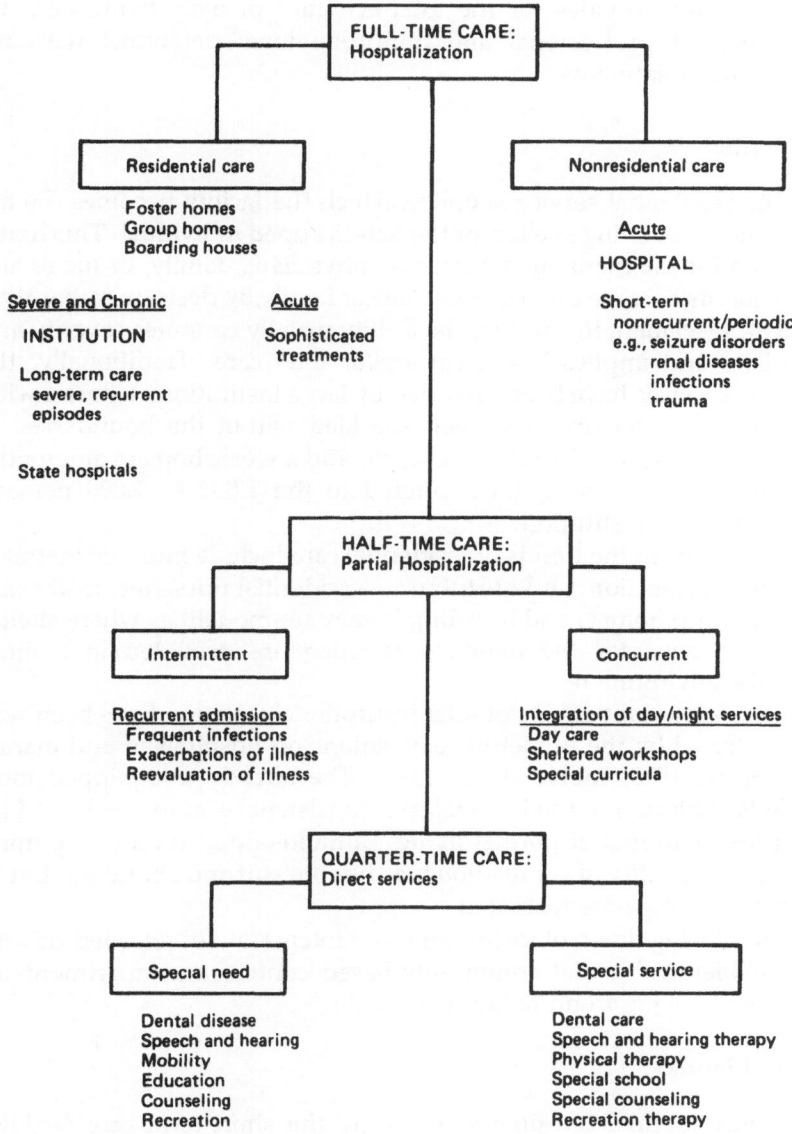

Figure 1. Hospitalization options from most to least restrictive.

Full-Time Care: Hospitalization

Full-time services include the total care setting and structured situations where the mentally retarded child or adult passes through the full course of the day within the confines of a facility. The full staff, including nurses, physicians, therapists, houseparents, attendants, or aides, assists in and provides for the total activities of daily living and the integration of professional and nonprofessional personnel with resources and treatments.

Residential

The residential service is one in which the facility becomes the assigned home or living shelter for the handicapped individual. This home is chosen for the mentally deficient by physicians, family, or judge and the residents become the person's nuclear family by decision rather than by birth. Placement in a residential facility usually connotes severity and chronicity and implies long-term and/or total care. Traditionally, the available services have been provided by large institutions where a wide spectrum of resources have been speckled within the boundaries. A hospital, a nursery, a laundry, a church, and a workshop are among the opportunities of a small town offered to the 1,000 to 4,000 persons confined in the institution.

Variations in the trends of residential care include multiple methods of shelter, protection, and nurturance. Residential nurseries, foster care homes, group homes, and boarding homes are modalities where shelter and nurturance for the mentally retarded are provided in a more homelike environment.

The deleterious effects of total institutional isolation have been well demonstrated by the numerous expositions on hospitalism and marasmus (Spitz, 1945; Ribble, 1939, 1944). The already unequipped individual is subjected in the hospital to inconsistencies of life-style and incongruous patterns of normal living. Stimulus opportunities are minimized; the quantity of stimulation may be constant and abundant, but its quality and range are restricted.

The closing down of institutions and integration of retarded citizens into smaller residential community-based centers and apartment arrangements is the trend today worldwide.

Nonresidential

Nonresidential full-time services are the short-term care facilities where the disabled child or adult is admitted for a defined and limited period of time; the institution or treatment center becomes the temporary shelter and protection agency. The length of stay is dictated by the nature of the admission and regulated by institutional authorities. There is no "indefinite" length of stay. The diagnostic rationale for admission becomes a prescriptive plan of care or cure for the individual.

The hospital usually serves as the locale for acute and short-term care services. The reasons for confinement are usually provided by the physician and predominantly dictated by a biomedical model. The environment is recognized as temporary, although activities of daily living are dominated by staff and schedules. Only recently have community resources, including educational programs, family living activities, and peer programs been employed for normalization of development and optimization of total health.

Half-Time Care: Partial Hospitalization

Half-time services provide an alternative to the total interruption of everyday living for the individual who needs supportive services not available in his usual living environment. This partial integration of therapies with the natural environment directs the necessary treatment at the specified need of the child or adult for a specific period of the day without entrenching the individual into a completely alien living environment.

These half-time arrangements are forms of partial hospitalization. The amount of time spent in the temporary therapeutic milieu can become equal to the amount of time spent in the child or adult's own home. The retarded child may be admitted to the hospital on a relatively constant and frequent periodic basis, or receive treatment or therapy on a specific area of need during some defined period of the day. This will be described as (1) intermittent and (2) concurrent partial hospitalization.

Intermittent

Intermittent hospitalization refers to a treatment model for a variety of biomedical and other dysfunctions. The reasons for admission are based on a medical model. For example, a child with multiple neurological problems might exhibit seizure activity to go along with subaverage mental ability. The seizures become a medical problem and are treated throughout the child's growth years with periodic readjustment of anticonvulsive medications. This often requires hospital admission and regulation of daily activities to monitor neurological dysfunction. The child is evaluated, adjusted, and sent back to a home environment in several days until reassessment is necessary again. During this time, the hospital staff cares for the child as ward conditions permit. The child with repeated admissions becomes a member of the ward "family," and readmissions are anticipated. The child must suffer the problems associated with life in several vascillating environments.

> Jane is a 4-year-old female presently enrolled in our mainstreamed preschool setting. She has a convulsive disorder stemming from a long medical cardiopulmonary history. She had heart surgery 2 years ago and at that time had a cardiac arrest. When her seizure activity was labile, she often was admitted to the hospital for convulsions with respiratory embarrassment.
>
> Jane also has a chronic respiratory problem secondary to a structural constriction of her lower trachea. She has continuous congestion in her lungs and is hospitalized every cold season for antibiotic therapy. The staff know her well and often imply that she has come "home."
>
> All in all, Jane has spent 2 of her 4 years of life in the hospital. An attempt to integrate services for her has resulted in that most of her hospitalizations have been arranged to be over school breaks in the past year. When this was not possible, she has been visited by school personnel in hospital who exchanged information with hospital staff to perpetuate some consistency in teaching Jane.
>
> Jane exhibited significant developmental delays when she was tested 1 year ago before starting school. She demonstrates clear improvement and developmental progress to date.

Concurrent

Concurrent partial hospitalization includes several different, integrated day and night programs provided for the retarded. It is the half-time service where the individual is provided treatment, not necessarily "cure"-focused, but rather within some context of natural day or night activity. Day care for the mentally retarded falls into this subdivision of partial hospitalization; children or adults are provided with activities that constitute their day, and return to their homes or residences at night.

The treatment program is usually individualized. The person may be scheduled into physical therapy, occupational therapy, vocational training, self-care activities, educational programming, or speech therapy, but the daytime appointments assume a normal day care schedule rather than a contrived hospital routine. The setting actually becomes part of the therapy.

Other partial services for the retarded might include such activities as sheltered workshops, work–study programs, or educational integration, particularly when the school curricula take on habilitation goals with engineered activities of daily living. The treatment itself is not problem-specific as is some therapy; however, a substantive integration of services exists over the entire course of the individual's day.

Part-Time Care: Direct Services

Direct services are the target-specific therapies provided for the mentally retarded. These activities are individual *and* problem-focused treatment: special service for special need. Dental care, speech and hearing treatments, recreation therapy, medical intervention, genetic counseling, family therapy, financial aid, and nutritional supplementary programs are among the direct services available for retarded citizens and their families.

Direct services represents the least restrictive and most normally integrated of intervention strategies. *All* services for mentally retarded individuals should be delivered within the philosophical context of normalization and with an awareness of the potential hazards of iatrogenic effects of treatment on development.

Guidelines for Hospitalization of Retarded Persons

We have derived several guidelines for use in considering and/or designing hospital programs for retarded children and adults. Adherence to these guidelines will minimize the negative impact health intervention might have on behavioral development as well as promote the least restrictive placement.

1. *Hospitalization should be developmental or, minimally, not counterdevelopmental.* Fundamentally, we refer here to the provision of stimulus and response opportunity. One would hardly argue that imprisonment of a young child or even an adult would be of any developmental benefit. However, what characterizes prison environments might also be found in certain institutions — including hospitals. Crowded wards, nonresponsive attendants, and little activity can create temporary and even long-term depressions in development.

Stimulus opportunities are most crucial, especially for the very young child. Sights, sounds, textures, smells — all avenues of stimulation — must be available *especially* for the retarded youngster who is already delayed. Periods of hospitalization where hours and even smaller durations are spent in relative stimulus impoverishment may be developmentally depressing. The hospital environment must be scrutinized and judged with respect to its range of available stimuli. The

totally quiet, "peaceful" ward may have administrative and staff
benefits but is not suitable for most children who are at risk in regard to
developmental regression.

Response opportunities become important as soon as the individual
is able to manipulate or move about in the environment. Objects and
aspects of the hospital environment should be available to permit active
involvement so that the person *can make things happen.* Custodial care
where the person remains a passive recipient of treatment is developmentally undesirable. Learning results from experiencing consequences
of behavior–environment interactions. The greater the scope and depth
of interaction, the more fertile is the developmental potential of the
environment.

The *contingencies of reinforcement* operating among stimuli, behavior,
and consequences are perhaps the most important considerations in the
developmental quality of the therapeutic situation. We refer here, of
course, to the functional relationship between behavior and preceding
or subsequent events. What is the *relationship* between a behavior and an
environment? What happens, for example, when a child screams while
lying in a hospital bed? Perhaps the consequence is staff attention, or
perhaps someone actually spends time playing with the youngster.
What happens when a child remains happily and quietly in bed? Frequently such desirable behavior is overlooked. The child's behavior
must deteriorate before it is reinforced! Partial hospitalization, because it
does not involve prolonged stays in the hospital, is less apt to foster
inadvertant counterproductive contingencies. But even short visits to a
setting where the contingencies are malevolent can undo school, home,
and clinic efforts to build positive development.

Because of the importance of the developmental environment, responsible individuals must examine and "grade" the therapeutic setting
in order to gain a rough idea of its possible benefit or harm to retarded
persons. Although it is only recently that researchers have turned to
screening and assessing the environment in addition to the person,
several guidelines and checklists are now available (Moos, 1974; Smith,
Neisworth, & Greer, 1978). These can be adapted to the hospital context
and may be quite useful in estimating the developmental potential of a
hospital environment and any program within it. In summary, we must
consider the effects on behavioral as well as biological development of
any hospital program. "Potential" for forward or retrograde development lies not solely within the patient but in the interaction between the
person and the program arrangements.

2. *Hospitalization should be curricular-based.* As much as possible,
school and hospital personnel should be aware of mutual objectives.
Often, activities in school can promote the therapeutic goals sought in

the hospital setting. Likewise, however, school-related objectives can be enhanced by a partial hospitalization program. Supplying certain current educational objectives to the hospital staff will assist them in promoting these objectives and in avoiding opposed objectives and procedures. School and hospital must cooperate to determine where conflict can be eliminated; coprofessional shaping and reinforcement of child progress can be effective. Perhaps even complementary progress in both biological and behavioral development may be achieved if a few developmental objectives are cooperatively written by school/hospital staff. Such interdisciplinary work is all too rare but is prerequisite to any real program that will articulate health and behavioral objectives and medical and educational procedures.

3. *Hospitalization should be normalizing or, minimally, not abnormalizing.* That is, therapeutic programs must not perpetually involve stigmatizing arrangements that draw attention to the person's exceptionality. Hospital identification devices (e.g., bracelets), specially marked parking spaces (e.g., "reserved for the handicapped"), special "sick clothing," and other earmarks of membership with a devalued group must be eliminated or sharply reduced.

As far as possible, retarded persons should "enjoy" no more special arrangements than nonretarded clients. Partial hospitalization does minimize much of the stigma associated with conventional hospital placement.

A well-designed partial hospitalization program can provide many of the health-maintenence benefits necessary while minimizing the visibility of exceptionality. Further, retarded individuals can be included along with nonretarded peers in aspects of a partial hospitalization program, such as group instructions, sitting together in a waiting room, eating lunch together, integrated play activities, and similar small group arrangements. Normalized provisions will offer normal models for retarded persons and reduce the social distance between retarded and nonretarded citizens. Partial hospitalization programs can thus be the health profession's expression of the normalization movement, as classroom mainstreaming has become for education.

References

Astrachan, B. M., Flynn, H. R., Geller, J. D., & Harvey, H. H. Systems approach to day hospitalization. *Archives of General Psychiatry*, 1970, *22*, 550–559.

Ayllon, T., & Michael, J. The psychiatric nurse as a behavioral engineer. *Journal of Experimental Analysis Behavior*, 1959, *2*, 323–334.

Beigel, A., & Feder, S. L. A night hospital program. *Hospital and Community Psychiatry*, 1970, *21*, 26–29. (a)

Beigel, A , & Feder, S L Patterns of utilization in partial hospitalization *American Journal of Psychiatry*, 1970, *126*, 101–108 (b)

Bijou, S W Practical implications of an interactional model of child development *Exceptional Children*, 1977, 44(1), 6–15

Blom, G E The reactions of hospitalized children to illness *Pediatrics*, 1959, *22*, 590

Guy, W , Gross, G M , Hogarty, G E , & Dennis, H A controlled evaluation of day hospital effectiveness *Archives of General Psychiatry*, 1969, *20*, 329–338

Moos, R H *Evaluating treatment environments A social ecological approach* New York Wiley, 1974

Neisworth, J T *Correlative biological and behavioral development* Talk presented at the Council for Exceptional Children, Atlanta, Georgia, 1977

Nirje, B The normalization principle and its human management implications In R Kugel & W Wolfensberger (Eds), *Changing patterns in residential services for the mentally retarded* Washington, D C President's Committee on Mental Retardation, 1969

Prugh, D G A study of the emotional responses of children and families to hospitalization and illness *American Journal of Orthopsychiatry*, 1953, *23*, 70

Report to the President, National Action to Combat Mental Retardation The President's Panel on Mental Retardation Washington, D C U S Government Printing Office, 1962, p 855 As cited in *The Pediatrician and the Child with Mental Retardation*, Committee on Children with Handicaps Evanston, Illinois, American Academy of Pediatrics, 1971

Ribble, M Significance of infantile sucking for psychic development *Journal of Nervous and Mental Diseases*, 1939, *90*, 455–463

Ribble, M A Infantile experience in relation to personality development In J McV Hunt (Ed), *Personality and the behavior disorders* New York Ronald Press, 1944 Pp 621–651

Smith, R M Neisworth, J T , & Greer, J G *Evaluating educational environments* Columbus, Ohio Charles Merrill, 1978

Spitz, R A Hospitalism An inquiry into the genesis of psychiatric conditions in early childhood A follow-up report *Psychoanalytic Study of the Child*, 1945, *1*, 53–74

Stuart, R B *Trick or treatment How and when psychotherapy fails* Champaign, Illinois Research Press, 1970

Wolfensberger, W *Normalization* Toronto National Institute on Mental Retardation, 1972

Wolman, B B *The therapist's handbook* New York Van Nostrand Reinhold, 1976

III

Research in Partial Hospitalization

Introduction

As in the case of nearly every aspect of contemporary psychiatric treatment, partial hospitalization is under pressure from various sources to substantiate the efficacy of its treatment efforts with empirical data. Until relatively recently, research in partial hospitalization has been sparse at best. Advocates of the treatment form seemed satisfied with establishing an apparent face validity for the modality through the construction of a theoretical or conceptual framework. Early data suggesting the therapeutic efficacy and cost-efficiency of partial treatment were accepted as final answers; little effort was expended to continue investigation and refine these preliminary conclusions.

In the last 5 years, however, renewed efforts have been made to expand the empirical base of partial hospitalization. This increase is reflected in the growing number of research reports appearing in professional literature as well as in a proportionate increase in research presentations at professional conferences.

In the following chapter, Michel Hersen presents a comprehensive, indeed, exhaustive, review of research in partial hospitalization; consideration is given to five separate areas of investigation and recommendations regarding research methods most applicable to the treatment modality are made; finally, suggestions for future areas of investigation and study are presented.

6

Research Considerations

Michel Hersen

Introduction

The initial impetus for partial hospitalization programming can best be traced to Cameron's (1947) paper in which he described an experimental day hospital at the Allan Memorial Institute of Psychiatry in Montreal, Canada. However, a more recent impetus for partial hospitalization programming can be identified. This is specifically concerned with the federal allocation of funds for mental health centers in general. As noted by Luber and Hersen (1976), with the implementation of the Mental Retardation Facilities and Community Mental Health Centers Construction Act of 1963 in the United States, there followed a 700% increase in day treatment programs and a 14% increase in the psychiatric population involved in such programs (see Taube, 1973). Indeed, in 1972 the population of new psychiatric patients treated in partial hospitalization programs in federally funded community mental health centers reached 37,117 (see Alcohol, Drug Abuse, and Mental Health Administration, 1974). Of course, this statistic does not include the thousands of patients treated in day and evening hospitals that were not receiving federal support for their operations.

As might be expected, concurrently with the increased number of patients being treated in partial hospitalization programs, many descriptions of the day-to-day activities carried out in such programs have

Michel Hersen • Department of Psychiatry, Western Psychiatric Institute and Clinic, University of Pittsburgh School of Medicine, Pittsburgh, Pennsylvania 15261.

appeared in the literature (e.g., Glasscote, Kraft, Glassman, & Jepson, 1969). When reading these descriptions, it is quite clear that the authors are dedicated to the notion that partial hospitalization can either supplant inpatient hospitalization altogether in some instances (thus providing a viable alternative) or that it can serve as a needed bridge between inpatient care and full community placement (see Hersen & Luber, 1977). In either case, the zeal and obvious enthusiasm of the proponents of partial hospitalization far outshadow the currently available data in support of their contentions. That is not to say that no data exist to support the viability of partial hospitalization programming. However, there are relatively few data-oriented papers in this area in contrast to the large numbers that have been written in which clinical descriptions have predominated. To illustrate the point, whereas there are several studies in which the relative efficacy of day hospitalization and inpatient hospitalization are contrasted (e.g., Herz, Endicott, Spitzer, & Mesnikoff, 1971; Michaux, Chelst, Foster, Pruim, & Dasinger, 1973; Washburn, Vannicelli, Longabaugh, & Scheff, 1976), there is only one study known to the present writer (i.e., Austin, Liberman, King, & DeRisi, 1976) in which two different approaches to partial hospitalization treatment (behavioral vs. eclectic) have been compared.

In spite of the aforementioned lacks, there are some empirically based studies that have appeared in the literature. For purposes of categorization, these data-based papers will be presented under five headings. The first will be specifically concerned with patterns of utilization (e.g., what kinds of diagnostic categories are to be found in day and evening hospital programs). The second deals with program evaluation. Here the studies evaluate within-program changes seen in patients over time. The third involves the comparison of partial hospitalization with inpatient hospitalization. The fourth is concerned with the one study where behavioral and eclectic procedures were contrasted in two separate day programs in two different hospitals. The fifth involves the use of the single-case research approach to problems presented by patients in the partial hospitalization setting. Here, research focuses on the individual patient, who serves as his/her own control while repeated measures of behaviors are taken during the course of experimental treatment.

In addition to providing a critical appraisal of the data-based papers subsumed under the five categories listed, a final section in this chapter will center on future directions that research might take in the partial hospitalization setting. Issues of record keeping and follow-up, and questions as to the administration of the appropriate treatment for the right patient are to be considered in some detail.

Patterns of Utilization

The papers to be discussed in this section are concerned with patterns of utilization seen in day and evening hospitals. A number of questions are to be raised in terms of (1) which diagnostic categories are admitted to programs, (2) what factors are predictive of complete utilization of partial hospitalization services, (3) what kinds of patients require overnight "guesting" on inpatient units, and (4) what are the attendance and participation factors for various activities and programs. Although several of the papers to be examined do not qualify strictly as illustrations of research, they do exemplify attempts on the part of proponents of partial hospitalization to document demographic data in a systematic fashion.

Hogarty, Dennis, Guy, and Gross (1968) have challenged the contention that the day hospital treatment center represents a viable alternative to inpatient hospitalization. This challenge is based on a study conducted in which 146 new admissions to the Baltimore Psychiatric Day Center (BPDC) (presumably a representative day hospital) were contrasted with a population of inpatients and outpatients. These 146 BPDC patients were carefully studied, using a variety of measurement instruments (Mental Status Schedule, Brief Psychiatric Rating Scale, Springfield Symptom Index, Katz Adjustment Scales, Hopkins Distress Index). These comprehensive assessments yielded the following picture:

> The characteristic day center admission is a white, married female with a high school education, living in the conjugal home with a spouse who is a skilled or semi-skilled employee. Unlike hospital admissions, nearly 80 percent of day center patients are in Social Class IV or higher as determined by the Hollingshead two-factor index of social class. Furthermore, of the 85 percent with a history of previous psychiatric treatment, more than 50 percent received psychotherapy alone or in combination with other treatments. Forty-six percent of patients have had prior hospitalizations but the number and length of hospitalizations are highly specific. Sixty-one percent of this group had but a single hospitalization, usually of less than a month's duration. (Hogarty et al., 1968, p. 936)

When the 146 BPDC patients were specifically compared with inpatient and outpatient samples, it appeared that the degree of psychopathology evidenced by day hospital patients fell somewhere between the inpatients and the outpatients. However, it was clear that BPDC patients were not at all representative of the types of patients usually found on an inpatient service:

> In summary, admissions to the BPDC appear to fall within an intermediate range of moderate to severe disorder. The selection process which affects referrals and/or admissions appears with regard to symptomatology as well.

> While the group presents significantly more disturbance than outpatients, there is notably *less* conceptual/perceptual disorder and *greater* social competence than is observed among inpatient admissions. Only among specific affective disturbances does the BPDC sample resemble inpatients. While the day center patients may be in need of hospitalization they appear to represent a subgroup of "potential" inpatients. (Hogarty *et al.*, 1968, p. 940)

Of course, the conclusions reached by Hogarty *et al.* (1968) are also subject to challenge, particularly with respect to the following question: Is the population of patients treated at the BPDC representative of patients being treated in other day hospital treatment centers? Indeed, in an earlier report (Craft, 1959), the diagnostic composition of six day hospitals in the United Kingdom, the USSR, the United States, and Nigeria was presented. Examination of these data suggests an extremely wide divergence of populations. For example, proportions of schizophrenics ranged from a low of 8% to a high of 47%. Manic-depressive psychosis ranged from 3% to 29% in the different hospitals. Similarly, anxiety neuroses ranged from a low of 9% to a high of 25%. To further illustrate the point, demographic data for a sample of male and female patients ($N = 100$) treated in a partial hospitalization service in a university hospital in the United States are presented in Table 1. Analyses of these data indicate that 63% of the male patients were schizophrenic; 53% of the female patients were schizophrenic. As noted by Hersen and Luber (1977), "we are dealing with a predominantly schizophrenic sample that has had numerous inpatient hospitalizations. With few exceptions, most patients were either unemployed or otherwise unproductive at the time of admission to our service. The majority were unmarried and socially isolated. Of course, these characteristics are typical of the chronic schizophrenic patient."

In consideration of the aforementioned, although the Hogarty *et al.* (1968) conclusions are of value in terms of the data reported, they should be restricted to the sample of patients considered. It is most apparent that there is tremendous variation in the diagnostic composition of partial hospitalization services. This probably relates to the unique referral sources, selection criteria, and philosophy of treatment that are peculiar to each day or evening hospital. Thus researchers should temper the generality of conclusions that may be reached on the basis of any one investigation.

Beigel and Feder (1970) identified those variables that were predictive of complete utilization of partial hospitalization services in a large New York City general hospital. In this study complete utilization was operationally defined as patients remaining in the day hospital program for at least 90 days followed by discharge to some kind of aftercare treatment. Incomplete utilization referred to patients who were transfer-

Table 1. Demographic Data for a Sample of 100 Patients Treated in the Partial Hospitalization Service[a]

Diagnoses	Median age	Median education	Employment at time of admission N	Marital status N	Median N of previous hospitalizations	Median length of stay (in days) in partial hospitalization service
(Male)						
Schizophrenia 31	24.8 yrs.	11.9 yrs.	Unemployed 41	Single 40	2.0	37.5 days
Unipolar depression 3	Range 16–76	Range 3–18	Employed 3	Married 2	Range 0–12	Range 2–120
Character disorders 4			Student 5	Separated 1		
Bipolar depression 2				Divorced 3		
Schizoaffective disorders 3				Widowed 1		
Neurosis 1						
Alcohol and drug abuse 2						
Mental retardation 3						15 patients still in program
N = 49						
(Female)						
Schizophrenia 27	32.0 yrs.	12.0 yrs.	Unemployed 47	Single 29	3.0	32.5 days
Unipolar depression 12	Range 18–61	Range 4–18	Employed 2	Married 7	Range 1–12	Range 2–111
Character disorders 2			Student 2	Separated 9		
Bipolar depression 3						
Schizoaffective disorders 4						
Neurosis 1						
Alcohol and drug abuse 2						17 patients still in program
N = 51						

[a]From Hersen and Luber (1977), Table 1.

red to an inpatient unit during the course of day hospitalization but who did not return to day hospital, those who prematurely left day hospital at their own initiative, and the group of patients that could not be managed in the day hospital. A variety of factors were evaluated with chi-square analyses for 176 patients. Diagnosis, suicide potential, and family involvement in treatment *were not* statistically significant predictors of complete utilization. However, when patients were divided into acute and chronic groups, the chronicity dimension proved to be of predictive value (see Table 2).

On the basis of this investigation, Beigel and Feder (1970) argue that two types of day hospital programs are needed to maximize treatment benefits for partial hospitalization patients. "The day hospital would provide short-term, intensive treatment designed as crisis intervention for the acute patient. The day care center would provide long-term treatment with an emphasis on social and vocational rehabilitation for the chronic patient. The availability of these two programs might significantly broaden the range of patients who could be effectively cared for in a partial hospital" (Beigel & Feder, 1970, p. 1273).

In a similar vein, Salzman, Strauss, Engle, and Kamins (1969) examined those variables that contribute to a day hospital patient requiring temporary full hospitalization on an inpatient unit (i.e., "overnight guesting"). During a 9-month period, 137 patients were admitted to the Day Hospital of the Massachusetts Mental Health Center, of which 20% required guesting at some point during their stay. Of the 28 patients who were guested, 11 were schizophrenic and 8 were depressives. Thus, severity of disturbance was greater for guested patients, with suicide potential and destructive behavior as the two primary indicants for the inpatient stay. Several other variables were also investigated (e.g., age, sex, marital status, parental social class, religion, family cooperation). Interestingly and contrary to expectation, patients who were guested did not come from "uncooperative families," but they did

Table 2. Diagnostic Comparison of Patients with Reference to Clinical State of Admission[a]

State	Complete utilization ($N = 115$)	Incomplete utilization ($N = 61$)
Acute ($N = 121$)	92	29
Chronic ($N = 55$)	23	32

[a]From Beigel and Feder. *American Journal of Psychiatry*, 1970, 126, 1269. Copyright 1970, American Psychiatric Association. Reprinted by permission.
$\chi^2 = 19.55$.
$p < .001$.

experience greater difficulty "disengaging from their families of origin or in establishing a lasting marriageable relationship."

Finally, the important issue of attendance and participation in the day hospital program has been tackled by Eckman (1976). The problem of attendance, in particular, is of considerable concern to all partial hospitalization programmers, inasmuch as few contingencies other than dismissal from the program are available in dealing with the recalcitrant patient (see Hersen & Luber, 1977). Even if attendance does not pose a major problem, there is no guarantee that the patient will actively participate in the scheduled activities. Of course, it is possible that many day and evening programs simply do not meet the needs and interests of the patients that they are presumably serving.

In attempting to combat this factor (i.e., patient apathy), Eckman (1976) describes a day hospital setting that emphasizes an educational workshop model. That is, instead of focusing on the hospital (i.e., the "disease" model), the program stresses reeducation and/or education by presenting a series of finite workshops that the patient must attend. Empirically based techniques and strategies emanating from the behavior modification framework are employed to foster the learning experience. Thus it is not uncommon for the "patient" to talk about homework and classroom assignments.

Eckman (1976) found that attendance at workshops (e.g., conversation skills, public agencies, personal finance) ranged from 63% to 94% (mean = 78.6%). Homework assignments were completed an average of 66.7% of the time, but classroom assignments were completed at a higher percentage (91.6%). Verbal participation in "classes" ranged from 72% to 98% (mean = 87.6%). Although no comparison data from other programs are currently available, the percentages reported here appear to be somewhat higher (i.e., in terms of participation) than is generally acknowledged in casual communication with day hospital directors. Of course, in the absence of hard data this statement remains conjecturable.

Program Evaluation

In this section the efficacy of a number of partial hospitalization programs will be evaluated. Although the data to be presented obviously cannot be described as representing controlled research, the conclusions derived therefrom *are* of some practical value to partial hospitalization programmers. However, even a simple statistical count of how many patients actually benefited in some way from attending a day or evening program is a definite improvement over impressionistic reports presented in many papers that have appeared in the literature. In general, the papers to be reviewed here contain data as to whether or not programs effected any changes in patients throughout their partial hospitalizations. Also, characteristics (e.g., diagnoses, other demographic variables) of patients who did and did not improve are identified.

Zwerling and Wilder (1964) evaluated the applicability of the day hospital in the treatment of acutely disturbed psychiatric patients. The particular program assessed was housed in a large municipal hospital center affiliated with a major medical school. The program was well staffed and involved pharmacological, individual, family, and group approaches to treatment. During an 18-month period, 189 patients were admitted from the same catchment area serving the hospital's inpatient unit. Of the 189 patients, 39% received their treatment entirely in the day hospital program. Five percent of the patients who were treated in the day hospital required guesting on the inpatient service for 1 or 2 nights; 22% required guesting on more than one occasion or for more than 2 nights. It might be noted that whereas 55% of the male patients were guested, the percentage for women was only 31. Zwerling and Wilder (1964) contend that "the sex difference in the need for transfer largely stems from the greater resistance and inability of families to cope at home with the aggressive threat of male patients" (p. 171). Thirty-four percent of the patients were rejected by the day hospital and treated on the inpatient services. Based on these results, the investigators concluded that day hospitalization offers a viable primary treatment alternative to inpatient hospitalization for some patients manifesting acute decompensation, but with a 24-hour hospital (inpatient unit) functioning in a backup capacity.

In a somewhat similar evaluation, Ruiz and Saiger (1972) documented the efficacy of a partial hospitalization program (serving an urban slum area in New York City) in circumventing admission to the local state hospital facility. In a 2-year period, 343 patients were initially admitted to an overnight inpatient unit (mean stay = 3.3 days) and then transferred to the partial hospitalization service as day patients. Approximately two-thirds of these patients were maintained in the community during their day hospital stay, thus apparently forestalling the more socially disruptive effects of complete hospitalization in the state facility. However, 31% of the patients did require transfer to the state hospital.

Lamb (1967), on a retrospective basis, examined the records of 51 patients admitted to the San Mateo County Psychiatric Day Hospital in California. One hundred and one patients who were admitted to the program (67%) successfully completed treatment (i.e., achieved goals of symptomatic reduction and social and/or vocational improvement prior to discharge). A further examination of successes versus failures in terms of chronicity and sex differences did not yield significant differences. However, chronicity of disorder was significantly correlated (.43) with length of stay on the unit. Average length of stay for chronic patients was 14.8 weeks; average length of stay for nonchronic patients was 8.7 weeks.

Chasin (1967) reports on 150 patients who were treated in the day hospital of the Massachusetts Mental Health Center, a well-staffed program affiliated with Harvard Medical School. Some of the data evaluated are presented with request to diagnostic groupings. From Table 3 it is clear that schizophrenics required more frequent guesting arrangements than any of the diagnostic categories. Schizophrenics were guested an average of 4.15 nights. Comparable averages for personality disorders and other patients in all categories were .87 and .03 nights, respectively.

The status of 121 day patients was also evaluated at follow-up. Analysis of these data (see Table 4) indicate that schizophrenics and personality disorders were less likely to be improved or working than patients in the other diagnostic categories (chi square $p < .01$). Thus it would appear that severity of disorder is inversely related to success in the day hospital program.

Johnson and Kish (1976) conducted a longitudinal study of patients at a Veterans Administration Day Treatment Center. Three assessment devices (Ellsworth MACC Behavior Adjustment Scale, the Personal Adjustment and Role Skills Scale, and the Veterans Adjustment Scale) were administered at 1-, 3-, and 6-month intervals following the patients' admission to the program. The results were analyzed with corre-

Table 3. Patients Transferred or Guested, by Diagnosis[a]

Diagnosis		Number of patients transferred	Number of patients guested	Number of nights of guesting
Schizophrenia, chronic	(N = 20)	2	2	44
Schizophrenia, chronic with acute exacerbation	(N = 13)	2	4	38
Schizophrenia, acute	(N = 21)	0	4	142
Affective psychosis	(N = 14)	0	0	0
Psychoneurosis	(N = 20)	0	1	1
Personality disorder	(N = 30)	3	4	26
Organic brain disease	(N = 3)	0	0	0

[a]From Chasin, Table 2. *American Journal of Psychiatry*, 1967, *123*, 782. Copyright 1967, American Psychiatric Association. Reprinted by permission.

Table 4. Status of 121 Day Patients at Follow-Up, by Diagnosis[a]

Diagnosis		Median span of hospital- ization (days)	Improved and working	Improved or working (but not both)	Not improved and not working
Schizophrenia, chronic	(N = 20)	74	4	6	10
Schizophrenia, chronic with acute exacerbation	(N = 13)	79	8	5	0
Schizophrenia, acute	(N = 21)	90	12	6	3
Affective psychosis	(N = 14)	45	7	7	0
Psychoneurosis	(N = 20)	58	10	9	1
Personality disorder	(N = 30)	39	13	6	11
Organic brain disease	(N = 3)	34	1	0	2

[a]From Chasin, Table 3. *American Journal of Psychiatry*, 1967, *123*, 784. Copyright 1967, American Psychiatric Association. Reprinted by permission.

lated t tests (differences between 1st and 3rd, 1st and 6th, and 3rd and 6th month). Of 75 such t ratios computed, only 5 reached statistical significance. Considering the fact that four t scores might have reached significance on a chance basis alone, the results are quite disappointing (i.e., *no change over time*). Of course, one problem with the study is the small number of patients actually assessed (ranged from 7 to 16). In commenting on their results, Johnson and Kish (1976) argue that at the very least the program followed the medical dictum of "not harming the patient." However, in this writer's judgment, this is not a sufficient justification; a treatment program can only justify its existence if demonstrable changes over time can be documented.

Sanders, Williamson, Akey, and Hollis (1976) offer a somewhat more optimistic report concerning the functioning of a behaviorally oriented intermediate day care center for adult psychiatric patients. The program was designed on a four-tier levels system, with patients advancing through the system when specific target behaviors within and outside of the program were achieved (see Table 5). Interrater agreement on *in-program* behavior was 91%; however, no reliability data were presented for self-rated behavior that occurred in the community. The investigators indicate that 79% of their patients advanced through the various levels of the program, but the exact number of patients actually treated is a bit difficult to ascertain from this report. Sanders *et al.* (1976) conclude that their preliminary study of this behavioral program "demonstrated ability to affect a change in patients' in- and outside-clinic problem behavior without the necessity for hospitalization. Although further evalua-

Table 5. Descriptions of Behaviors Defined at Each Achievement Level[a]

		Behavioral categories		
Level	Social skills	Problem-solving group	Work duties	Specific behavior[b]
1	Be on time Call if absent Ask to leave Knock on doors Participate Eat without making a mess No alcohol/drugs	Talk about your problems co- herently Look at/listen to others	Kitchen clean-up Household clean-up Yard clean-up	In clinic: Talk to 2 people for 5 minutes each within 30 minutes Outside-clinic: Baseline — count conversations ini- tiated
2	Same as above	Talk about your specific behavior problems on card Look at/listen to others	Same as above	In clinic: Talk to 3 people for 10 minutes each within 30 minutes Make 2 positive statements about yourself each day Outside clinic: Initiate 1 conver- sation each day
3	Same as above	Give 1 solution to a problem Give 2 others ideas for solving their problems Look at/listen to others	Same as above	In clinic: Lead a 30-minute discus- sion with 2 people Make 4 positive statements about yourself each day Outside clinic: Initiate 2 conver- sations each day
4	Same as above	Tell group results of problem-solving actions Give appropriate praise to 3 others Give 3 others ideas for solving their problems Look at/listen to others	Same as above	In clinic: Same as above Make 8 positive statements about yourself each day Lead all exercise groups Outside clinic: Initiate 4 conver- sations each day

[a]From Sanders, Williamson, Akey, and Hollis (1976), Table 1.
[b]Specific behavior listed is that of a single client, taken as an example of typical behavior defined.

tive data and systematic revisions are obviously needed, the program's comparative effectiveness is encouraging and offers promise for the future" (pp. 281–282).

Comparison of Partial Hospitalization with Inpatient Hospitalization

In order to rebut the critics of the partial hospitalization concept, it behooves the proponents of partial hospitalization to demonstrate the effectiveness of their programs on a comparative basis. Of course, the standard for comparison will undoubtedly be the inpatient unit, particularly if partial hospitalization is to be considered a viable alternative to inpatient placement for a select group of patients. As noted in the introductory section to this chapter, there are several studies in which day hospital programs and inpatient programs have been directly contrasted. In this section the results of four such studies will be considered in some detail.

Wilder, Levin, and Zwerling (1966) describe a 2-year follow-up study of 189 patients initially admitted to an inpatient unit and 189 patients initially admitted to a day hospital program. Assignment to the two treatment modalities was done randomly after the "admitting doctor" had decided inpatient treatment was required. In either case, the admitting physician was unaware of the random assignment procedure, thus avoiding possible bias in the selection procedure. Although the respective staffs of the day and inpatient services had comparable training and experience, there were some distinctive differences in the settings. The day hospital program involved its patients in individual, family, and group psychotherapy, in addition to the ward "therapeutic community" orientation. On the other hand, the inpatient unit followed "the more traditional occupational therapy and recreational therapy models." (It seems as though the inpatient unit fostered a somewhat custodial orientation, albeit probably unintentionally.) It should be noted that the day hospital rejected one-third of the patients admitted for hospitalization. In half of the cases rejected, the primary diagnosis was acute or chronic brain syndrome.

Hospitalization and rehospitalization durations for the two groups are presented in Table 6. Examination of the table indicates that initial treatment in the day hospital took significantly longer than the inpatient service. This was probably due to the pressures for maintaining the consistent flow of patients admitted to the inpatient unit. However, the interval between the project discharge date and the first readmission

Table 6. Hospitalization and Rehospitalization Durations and Intervals (Median Days)[a]

Duration[b]		Day hospital	Inpatient service
Project hospitalization duration			
Day hospital	$N = 138$		
Inpatient service	$N = 138$	57	20[c]
Interval between project discharge and (first) readmission			
Day hospital	$N = 56$		
Inpatient service	$N = 61$	271	162[c]
Interval between project admission and (first) readmission			
Day hospital	$N = 56$		
Inpatient service	$N = 61$	318	247
Cumulative rehospitalized days during the 2-year period			
Day hospital	$N = 56$		
Inpatient service	$N = 61$	113	74
Cumulative hospitalized days during the 2-year period			
Day hospital	$N = 137$		
Inpatient service	$N = 135$	78	64

[a]From Wilder, Levin, and Zwerling, Table 2. *American Journal of Psychiatry*, 1966, 122, 1098. Copyright 1966, American Psychiatric Association. Reprinted by permission.
[b]All durations are charted as medians since the variability of duration indices in our sample was so great as to rule out the mean as a reliable parameter.
[c]The difference between the day hospital and the inpatient service is significant at $p < .01$.

was significantly longer in the case of the day hospital. All other comparisons indicated in Table 6 are not significant.

Comparisons of patient psychiatric status (rated by patients and family members) for the 2-year project did not yield significant differences. However, family adjustment ratings (as assessed by the patient) were significantly higher for patients treated in the inpatient unit. Interestingly, this difference was not significant when family member ratings on this same dimension were contrasted for the two programs.

Wilder *et al.* (1966) also analyzed their outcome data in terms of sex and diagnosis. For example, for both men and women with a diagnosis of affective psychosis, there was a significantly longer interval between admission and readmission for day hospital patients than for inpatients. Schizophrenic women admitted to day hospital as opposed to the inpatient service seemed to do better in their posthospital adjustment. That is, at follow-up, 93% of day hospital schizophrenic women were in the community; only 69% of inpatient schizophrenic women were in the community. This difference was statistically significant and can be accounted for in part by the higher readmission rate of the inpatients.

In general, the findings of this study support the contention that day hospital can be an effective alternative to inpatient care for a large percentage of acutely disturbed psychiatric patients. Indeed, Wilder *et al.* (1966) conclude that "the day hospital was a feasible treatment modality and was generally as effective as the inpatient service in the treatment of acutely disturbed patients for most or all phases of their hospitalization. It was sobering to note that the clinical course for many of the patients in both treatment groups was marked by considerable disability and frequent hospitalizations. The aftercare of these patients was minimal and may have contributed to these findings" (p. 1101).

A most interesting comparison of day versus inpatient hospitalization was conducted by Herz *et al.* (1971) in a hospital associated with the New York State Psychiatric Institute. Probably the most unusual feature of the study was the fact that the same treatment facility and its staff served both the inpatients and the day hospital patients. Thus the problem of matching staff and facilities was obviated. The program featured the full range of psychiatric services including psychotherapeutic and chemotherapeutic programs. Unfortunately, however, day hospital patients and inpatients were imperfectly matched such that inpatients were younger, had a greater proportion of schizophrenics, and were of a higher social class. Nonetheless, initial differences between groups were considered in the analysis of resulting data.

Prior to random assignment to groups, patients spent an average of 3 days on the inpatient unit. Over a 2-year period all new admissions to the inpatient unit were screened (total $N = 424$ patients). Of these patients, 22% (90 patients) were accepted into the protocol study (45 patients in each group). Most of the rejections were due to their being either "too psychiatrically healthy or too psychiatrically ill." Psychopathology and role functioning of patients were evaluated using the Psychiatric Status Schedule and the Psychiatric Evaluation Form.

The results of this study are quite important inasmuch as they favor the day hospital program on practically each dependent measure assessed. To begin with, for patients discharged into the community who stayed there at least 1 week following discharge, average length of stay on the inpatient service was 138.8 days as contrasted to 48.5 days for partial hospitalization patients. At the 4-week follow-up period, day patients showed significantly greater improvement than inpatients on 5 of 19 scales on the Psychiatric Evaluation Form. However, it might be pointed out that the investigators accepted significant differences using a .10 level (two-tailed) rather than the more universally accepted .05

level. At the long-term follow-up evaluation, day patients showed superior improvement over inpatients on the Daily Routine–Leisure Time Impairment, and Housekeeper–Role Impairment Scales.

In attempting to account for the superior performance of day hospital patients, the investigators speculate as follows:

> Day care apparently avoids the regressive feature associated with "total institutionalization." In addition, day patients have a greater opportunity to maintain healthy areas of functioning, including the preservation of social and instrumental roles. Another major factor is that the powerful therapeutic effects of psychotropic drugs probably make it unnecessary to subject the patient to the stress of complete separation from familiar ties in order to effect a remission in the illness. (Herz et al., 1971, p. 1379)

However, the specific therapeutic role of pharmacological interventions was not contrasted in this study and may represent a potential confound if differential drug regimes were given to patients in the two programs.

Michaux et al. (1973) compared adjustment of day hospital and full-time psychiatric patients at 2 and 12 months postdischarge. Of 50 day patients and 56 inpatients who completed treatment, 45 of the day patients and 52 of the inpatients were available for postdischarge evaluations. Several measurement instruments were used to assess the respective efficacy of the two treatment modalities (Multidimensional Psychiatric Scale, Katz Adjustment Scale, Michaux Stress Index, and overall ratings of "improvement, severity of illness, adaptability, and discomfort"). These evaluations were completed at stated intervals by an assessment team that was not involved in either of the two treatments.

Examination of aftercare afforded both groups indicates that it was similar in terms of number of contacts, medication and/or its discontinuation, and relapse rates. Although at the time of discharge from their respective programs significantly greater symptomatic improvement had been observed in the inpatients, these differences disappeared over time. Specifically, at the 2-month follow-up evaluation, differences between the two treatment groups were less marked. However, day patients were still "more conceptually disorganized, hyperactive, and paranoid than their full-time counterparts." On the other hand, at the 1-year follow-up the only major difference was that day patients were significantly more intrapunitive than the inpatients.

By contrast to the aforementioned, at the 1-year follow-up, improvements in social functioning were very much in favor of the day hospital patients. Of the patients who were "heads of households," 78% of the day patients were working but only 52% of the inpatients were employed. In general, the results point to the differential effectiveness of the two modes of treatment. As stated by Michaux et al. (1973):

This study revealed special merits and limitations of both day treatment and full hospitalization — the latter providing quick symptomatic relief, the former enabling patients to make lasting gains in social adjustment and eventually in symptomatology. These findings should, hopefully, contribute to a clearer definition of the treatment goals of both day-care center and hospital and to a more thorough understanding of the function of each in a comprehensive mental health care program. (p. 651)

A most recent comparative study of day hospitalization and inpatient hospitalization appears in a report by Washburn *et al.* (1976). Following 2 to 6 weeks of inpatient assessment, 30 seriously disturbed female patients were given additional inpatient care. Twenty-nine equally disturbed female patients who initially underwent the extended inpatient evaluation were assigned to day hospital treatment. In addition, 34 female patients, already in the day center program, served as a control group. Most of the 93 patients followed in the study were middle class, with 50% diagnosed schizophrenic, 12% affective psychosis, 20% personality disorder, and 18% borderline personality. Patients in the three conditions were evaluated over a 2-year period on several dimensions (global mental status: subject form; global mental status: informant form; global mental status: psychiatric evaluation form; impulse control; community adjustment; family adjustment: informant form; family adjustment: subject form; intrapsychic functioning; number of roles attempted; burden evaluation; direct charges; and days of attachment).

The results of this study favored the day hospital group on five dimensions: (1) subjective distress, (2) community functioning, (3) family burden, (4) total hospital cost, and (5) days of attachment to the program. However, it was found that at the 24-month assessment the general superiority of the day hospital group was no longer in evidence. That is, the inpatients had "caught up" and evidenced comparable gains. Notwithstanding this caveat, the importance of day hospitalization as a viable primary treatment modality is underscored. But, given the obvious advantages of day hospitalization over total inpatient hospitalization, why is there such widespread *underutilization* of day hospital as a major treatment approach to severe psychiatric disorder? Other than financial considerations (i.e., greater reimbursement by insurance carriers for inpatient utilization), the following conclusions were reached by Washburn *et al.* (1976):

We suggest, however, that the main factor leading to underutilization of day hospitals is the attitude of primary treatment personnel. The primary therapist feels the responsibility and pressures of the patient's unpredictable behavior and, for peace of mind, seeks the control of 24-hour inpatient hospitalization. To comfortably use the day treatment option, the therapist needs additional emotional support and professional approval. Such support can be provided by increasing front-line staff in day settings and by dis-

seminating studies such as the present one, which indicates that many patients (even most) are best treated while residing in the community. We
anticipate that the analysis of our process data will be especially useful in
providing patient profiles of those most likely to succeed in day treatment.
(p. 675)

To summarize, the results of the aforementioned studies clearly support the contention that the day hospital provides a viable alternative to full inpatient hospitalization. Although all patients admitted to the inpatient services may not be suitable for day hospital care, the percentage is sufficiently large such that better utilization of partial hospitalization programming is definitely warranted. The studies reviewed provide evidence that the patient who is taking part in a day hospital program is much more likely than his/her fully hospitalized counterpart to evince appropriate role functioning in the community at given assessment intervals. Further, even though inpatient hospitalization may initially result in a significantly greater decrease in symptomatology than in the case of partial hospitalization, over time these initial differences tend to dissipate. Thus a very good case has been made in favor of the partial hospitalization concept.

Behavioral and Eclectic Procedures Contrasted

It is of interest (but unfortunate) that in the three decades since Cameron's (1947) early description of the day hospital, there has been only one study in which different styles of partial hospitalization programming have been contrasted. Thus, despite any criticism that may be leveled at the work of Austin *et al.* (1976), these investigators should be applauded for engaging in the necessary first step. In the Austin *et al.* (1976) investigation, the comparative efficacy of two day hospital programs situated in mental health centers in two relatively similar catchment areas (respective populations = 140,000 and 120,000) was studied. Although the composition of the staff of the two day treatment centers was quite similar, the treatment orientation in each presumably was different. The Oxnard Day Treatment Center (Oxnard DTC) followed the behavioral-educational approach. By contrast, the Day Treatment Center II (DTC II) was described as "eclectic" and "milieu-oriented."

Thirty patients from the Oxnard DTC and 26 patients from the DTC II served as subjects for this study. (See Table 7 for a description of the characteristics of the two samples.) Examination of the table shows that despite some obvious differences on a number of the variables, the patients were relatively well matched on most.

Table 7. Background Characteristics of Patients at Oxnard DTC and DTC II Who Were Evaluated by Goal Attainment Scaling[a]

Characteristics	Oxnard DTC (N = 30)	DTC II (N = 26)
	%	%
Diagnosis[b]		
Schizophrenic	40.0	34.6
Manic-depressive	13.3	7.7
Neurotic (anxiety and depression)	36.7	46.2
Personality disorder	10.0	3.8
Substance abuse	0	7.7
Sex		
Male	43.3	46.2
Female	56.7	53.8
Education		
College	30.0	23.0
High school graduate	36.7	46.2
Some high school	26.7	23.1
No high school	6.6	7.7
Age		
16–30	53.3	61.5
31–40	16.7	15.5
41 and above	30.0	23.0
Marital status		
Single	53.3	57.7
Married	36.7	23.1
Divorced and separated	6.6	11.5
Widowed	3.4	7.7
Income		
Dependent (government support, family support)	86.7	80.8
Self-supporting	13.3	19.2
Previous hospitalization		
Yes	53.3	73.0
No	46.7	27.0
Median attendance		
Days in day treatment program	39.0	27.5

[a]From Austin, Liberman, King, and DeRisi (1976), Table 1.
[b]Diagnoses were made by the clinical staff at the time of intake.

In assessing differences between the two programs, Austin *et al.* (1976) decided against using recidivism rates (a most typically employed dependent measure) as the main dependent variable. Instead, the investigators chose to follow a procedure known as goal attaining scaling. This was defined as follows: "The status of each patient with regard to his/her goals was indicated at intake and at follow-up by the rater checking the scale level which best represented the patient's current functioning. The significant other supplied external validation of the patient's goals and progress at follow-up. Future projections of less and least favorable outcomes were most often marked as the starting points or status at intake for patients who, it was felt, would reach the expected level of treatment success or better. Treatment outcome is measured by movement from one scale level to another and by converting the scale levels reached at follow-up to a standardized Goal Attainment T-score" (Austin *et al.*, 1976, p. 256).

The results of this study are presented in Table 8. Examination of this table indicates that percentile ranks for the Oxnard DTC were higher at each of the follow-up measurement points (3, 6, 24 months). However, with the exception of the 24-month follow-up comparison, which only approached significance ($p. < .10$), none of the remaining differences was statistically significant. A further evaluation of the two programs was made after it was discovered that one of the therapists in the DTC II (ostensibly the nonbehavioral program) had been using behavior therapy strategies with her patients. Thus the data were reanalyzed minus this therapist's patients. On the basis of a reanalysis of the data, and additional statistically significant difference in favor of the Oxnard DTC was found at the 6-month follow-up assessment.

In summary, both treatment programs seemed to be effective, with a slight superiority for the behaviorally oriented approach. However, this superiority emerged after many statistical convolutions were performed. It it clear, then, that considerably more research of the kind

Table 8. Mean Goal Attainment Scores for Oxnard DTC and DTC II Patients at Intake, 3-Month, 6-Month, and 24-Month Follow-Up Points[a]

	Intake level	3-Month follow-up	6-Month follow-up	24-month follow-up
Oxnard DTC				
T-score	25.60	60.92	64.69	66.43
Percentile rank	.6	86.00	93.00	94.50
Standard deviation		12.43	10.85	13.45
$N =$	30	28	27	18
DTC II				
T-score	25.29	57.88	58.12	55.82
Percentile rank	.4	79.00	79.00	73.00
Standard deviation		13.70	18.38	23.22
$N =$	26	19	23	12

[a]From: Austin, Liberman, King, and DeRisi (1976), Table 4.

reported by Austin *et al.* (1976) needs to be conducted before any conclu-
sive statements can be made about the relative efficacy of various day
treatment approaches that are currently in vogue in the field. Also, it
would appear to be of considerable import to monitor therapist behavior
in programs where comparisons are to be undertaken in order to ensure
that treatments are actually accorded consistent with the investigator's
stated and written protocol.

Single-Case Research Approach

The single-case approach to research is ideally suited to the partial
hospitalization setting. Although this research strategy has been widely
employed in psychiatric, educational, and rehabilitation settings (see
Barlow & Hersen, 1973; Hersen & Barlow, 1976, for reviews), it has been
infrequently applied in either day or evening programs (exceptions include: Bellack, Hersen, & Turner, 1976; Hersen & Bellack, 1976a; Liberman & Smith, 1972; Lombardo & Turner, in press; Turner, Hersen, &
Bellack, 1977). This is somewhat surprising inasmuch as the relative
stability of the partial hospitalization population (i.e., their repeated
availability on a day-to-day basis) and the rather lengthy stays in particular programs are both highly compatible with the requisites of this
research strategy.

Although, as noted above, the single-case design has been described at length (see Hersen & Barlow, 1976, for a complete exposition),
it might be useful to briefly review some of the basic tenets of the basic
approach. This, then, will be followed by presentation of published
illustrations of some of the more widely used single-case design tactics.

First, in single-case research, the patient serves as his or her own
control. That is, rather than being assigned to an experimental treatment, a "placebo" treatment, or a waiting list control, the patient may be
exposed to alternating baseline and treatment phases over time. Thus
repeated measures of the patient's behavior(s) are obtained. These may
vary from observed motoric responses, self-reports, or physiological indices, each being assessed under standardized conditions. By obtaining
repeated measurements over time in such fashion, the experimenter-
therapist is able to evaluate the progress (or lack thereof) of the treatment being administered. This permits flexibility in the research strategy
(a feature uncommon to group comparison designs) in that ineffective or
inefficient treatments can be replaced at will or modified until desired
effects are achieved.

In the basic A-B-A-B design, baseline (A) and treatment (B) are
alternated, allowing for analysis of the *controlling* effects (over the de-

pendent measure) of the specific treatment. For example, after baseline (A) measurement is established (i.e., the natural frequency or rate of the behavior under study), treatment (B) is applied. Resulting data may indicate that treatment is yielding an improvement in the behavior. Then, treatment will be withdrawn and the second baseline (A) phase follows. When treatment is withdrawn a decrement in the behavior should ensue. Should this be the case, the controlling effects of the treatment over the dependent measure will have been established. At this point treatment is reinstituted (second B phase), and improvements should once again be noted. If this occurs, the controlling effects of the treatment (A to B) will have been established yet a second time. It should be pointed out that many variants of the basic A-B-A-B design have been used by behavioral researchers (see Hersen & Barlow, 1976).

Another basic strategy followed by single-case researchers is the multiple-baseline design. Here, several specific behaviors for a given individual patient are targeted for treatment. Baselines for each are obtained simultaneously (presumably the behaviors selected are independent of one another) and then treatment is applied to the first. After the first treated behavior has reached the preset criterion, treatment is applied successively and cumulatively under time-lagged conditions to the remaining behaviors. Resulting data are analyzed visually by inferring changes in the treated baselines from those that remain untreated. When the baselines are independent, change occurs only as the treatment is directly applied to each. However, if change should take place concurrently in an untreated baseline, the independence of behaviors will not have been established, making firm conclusions as to the controlling effects of the treatment difficult to render. This design is particularly useful when the A-B-A-B type strategy is not feasible (e.g., when treatment cannot be withdrawn, as in the case of instructions or due to ethical considerations.

Lombardo and Turner (in press) used an A-B-A-B design to assess the effects of thought stopping in controlling obsessive ruminations in a chronic schizophrenic patient. The patient was 25 years old and had a long history of prior hospitalizations, beginning when he was 16. Despite receiving psychotropic medication, the patient spent many hours each day obsession about "fantasied heterosexual relationships."

Throughout all phases of the study the patient participated in the regularly scheduled group activities in the Day Hospital at Western Psychiatric Institute and Clinic, University of Pittsburgh School of Medicine. During baseline assessment and treatment phases he was instructed to record beginning and end times of each obsessive rumination on a specific form provided by the therapist. After a 6-day baseline period, thought-stopping (see Wolpe, 1973) treatment was begun. This involved daily 75-minute sessions in which obsessive ruminations were interrupted by the therapist shouting, "Stop," contingent upon the patient signaling his therapist that he was obsessing. To facilitate the patient achieving self-control over the behavior, he was taught to verbalize the word *stop* contingent on ruminations; he then was taught to do this covertly. After 10 days of treatment, thought-stopping procedures were withdrawn for 10 days (second baseline phase) and then reinstated for 13 days (second treatment phase).

The results of this study appear in Figure 1. Examination of the data indicates a high rate of ruminations during the first baseline, a substantial decrease during thought stopping, a marked increase in the second baseline when treatment was discontinued, and a considerable improvement leading to a zero rate when thought stopping was reinstituted. Moreover, follow-up data collected for 6 weeks posttreatment indicated maintenance of improvement. In short, this experimental analysis shows that thought stopping was an effective and controlling variable in reducing obsessiveness in this schizophrenic patient.

A more extended variant of the basic A-B-A-B design (i.e., A-B-A-C-A-C-A-D-A) was used by Turner *et al.* (1977) in evaluating the effects of social disruption, stimulus interference, and aversive conditioning for auditory hallucinations in a chronic schizophrenic patient. The patient was a 33-year-old female, with an 11-year history of her

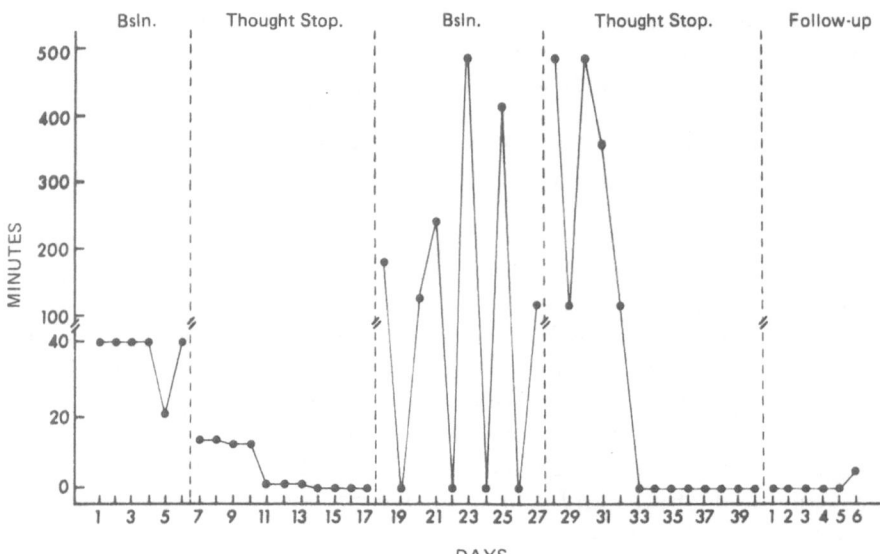

Figure 1. Duration of obsessive rumination during baseline treatment, and 6-week follow-up. From Lombardo and Turner (in press), Figure 1.

disorder necessitating 10 hospitalizations. During the course of the study she was a patient in the Day Hospital at Western Psychiatric Institute and Clinic. Despite her being medicated with Haldol 5 mg, q.i.d. and Prolixin Decanoate 25 mg, I.M., she complained of hearing persistent voices instructing her to engage in a variety of unpleasant actions (e.g., "kill your roommate"). Therefore, several behavioral strategies were implemented in sequence in order to help her control the rate of auditory hallucinations.

The experimental analysis consisted of 10 phases: (1) *Baseline 1.* In this phase percent time hallucinating and frequency of hallucinations were assessed. The patient sat in a comfortable chair and was instructed to raise her right index finger when she heard the "voice," keeping the finger raised until the "voice" disappeared. Assessment sessions were of 20 minutes' duration and were held twice daily. The patient was observed by a research technician while a second research technician provided the reliability check for one-third of the sessions. (2) *Social Disruption.* In this phase the patient continued signaling about hallucinations, but the research technician involved her in non-sympton-related conversation. (3) *Baseline 2.* (4) *Social Disruption.* (5) *Stimulus Bell.* In this phase an electrically operated bell (70 decibels) was activated upon the patient signaling hallucinations. (6) *Baseline 3.* (7) *Stimulus Bell.* (8) *Baseline 4.* (9) *Electric Shock.* In this phase faradic shock ranging from 6 to 10 milliamperes was administered to the patient (electrodes attached to her left wrist) whenever the appearance of hallucinations was signaled. If hallucinations persisted the shock was maintained until the patient signaled that the "voice" had terminated. (10) *Baseline 5.*

The results of the study are presented in Figure 2 and indicate that the bell in phases 5 and 7 resulted in the most substantial diminution in percentage time and mean duration engaged in hallucinatory behavior. However, the results were ephemeral as indicated by renewed increases in succeeding baseline phases. Also, the effect of the bell on frequency of hallucinations was minimal. Social disruption (phases 2 and 4) similarly resulted in decreased percentage time and mean duration hallucinating. Indeed, it appears that both social disruption and the stimulus bell served only to *distract* the patient from her hallucinations. On the other hand, application of faradic shock in phase 9 appeared to effect changes in percentage time hallucinating, mean duration of hallucinations, and frequency of hallucinations to some extent. Improvements noted in phases 9 and 10 were seen in the follow-up assessments 22 and 25 weeks posttreatment.

A multiple-baseline strategy was followed by Hersen and Bellack (1976a) in their assessment of social skills training in a chronic schizophrenic. The patient was a 19-year-old schizophrenic whose withdrawal

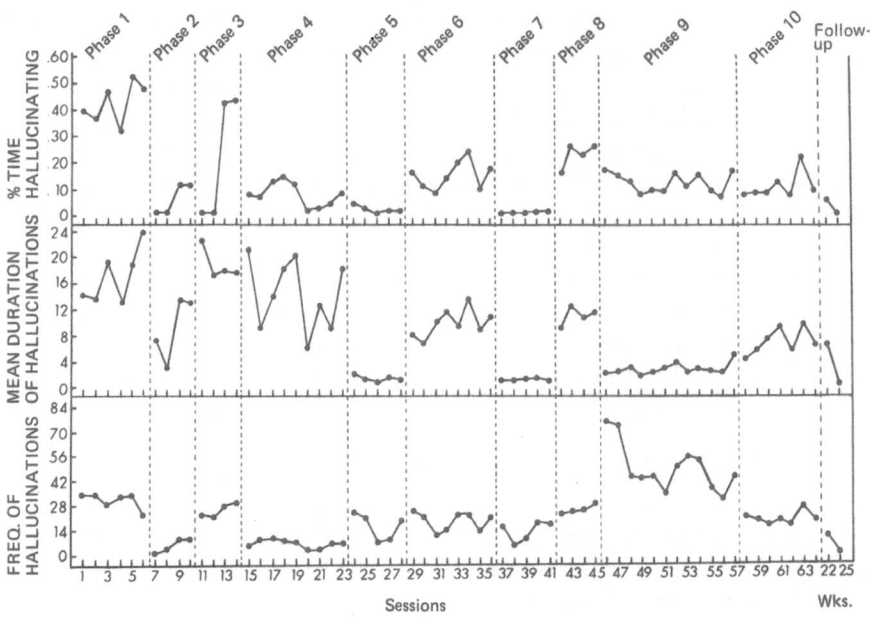

Figure 2. Percent time, mean duration, and frequency of hallucinatory behavior during probe sessions and follow-up in a chronic schizophrenic treatment. From Turner, Hersen, and Bellack (1977), Figure 1.

and psychotic symptomatology had persisted for 4 years. During his stay in the Day Hospital at Western Psychiatric Institute and Clinic, the patient was given Trilafon 32 mg H.S. and Cogentin 2 mg, H.S.

An initial baseline assessment of the patient in role-played scenes (BAT-R) (see Eisler, Hersen, Miller, & Blanchard, 1975) that were videotaped indicated skill deficits in several areas: poor eye contact, short speech duration, inability to make requests of an interpersonal partner, extensive compliance with requests of others, and general unassertiveness. Subsequent to baseline assessment, the patient received five 20- to 40-minute social skills training sessions per week for each targeted behavior. Social skills training here consisted of instructions, feedback, and behavioral rehearsal, using four scenes requiring positive assertion and four scenes requiring negative assertion taken from the BAT-R. Such training was applied to each behavior under time-lagged conditions.

The results of the experimental analysis are presented in Figure 3 and indicate that for the first three behaviors treated, improvements occurred only when social skills training was directly applied to each. However, as treatment was applied to requests, number of compliances decreased concurrently, indicating the nonindependent nature of these

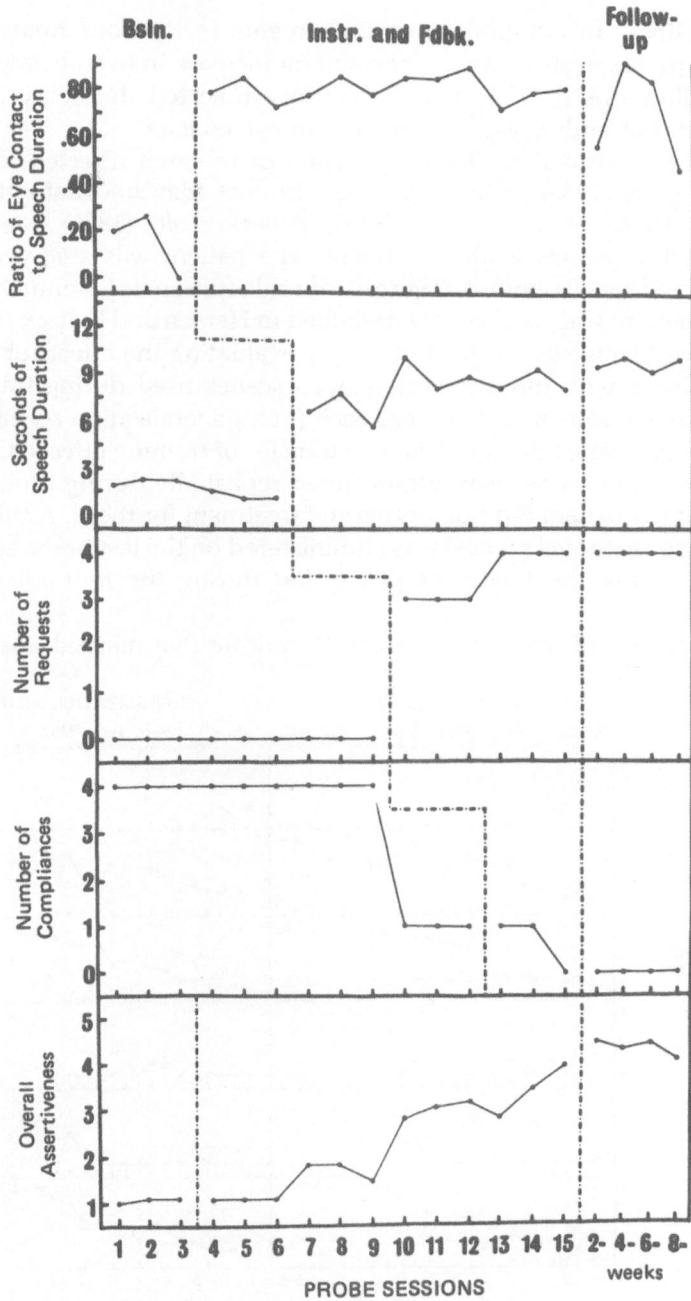

Figure 3. Probe sessions during baseline treatment, and follow-ups for Subject 1. Data are presented in blocks of eight scenes. From Hersen and Bellack (1976a), Figure 1.

two baselines. In addition, as improvements in the four treated behaviors appeared, there was a concomitant increase in overall assertiveness. Follow-ups at 2, 4, 6, and 8 weeks indicated durability of the
treatment, but with a slight decrement in eye contact.

In the continuation of their program of research directed toward
improving social skill deficits in schizophrenics (day hospital patients)
(see Hersen & Bellack, 1976b, 1977), Bellack *et al.* (1976) examined
generalization effects of the treatment. The patient was a 48-year-old
unemployed female with a diagnosis of schizophrenia. Essentially, the
same paradigm was used as that described in Hersen and Bellack (1976a)
(see above). However, instead of simply evaluating the effects of social
skills training with the eight role-played scenes used during training,
two additional sets of eight scenes each (i.e., generalization scenes and
novel scenes) were added to evaluate transfer of training effects. One of
the sets of eight scenes was administered repeatedly during probe sessions, but the patient did not receive any treatment for these. A third set
of eight scenes (novel scenes) was administered on the last probe session
(no. 17) during the treatment phase and during the four follow-up
probes.

The results of this study (Figure 4) indicate that marked improve-

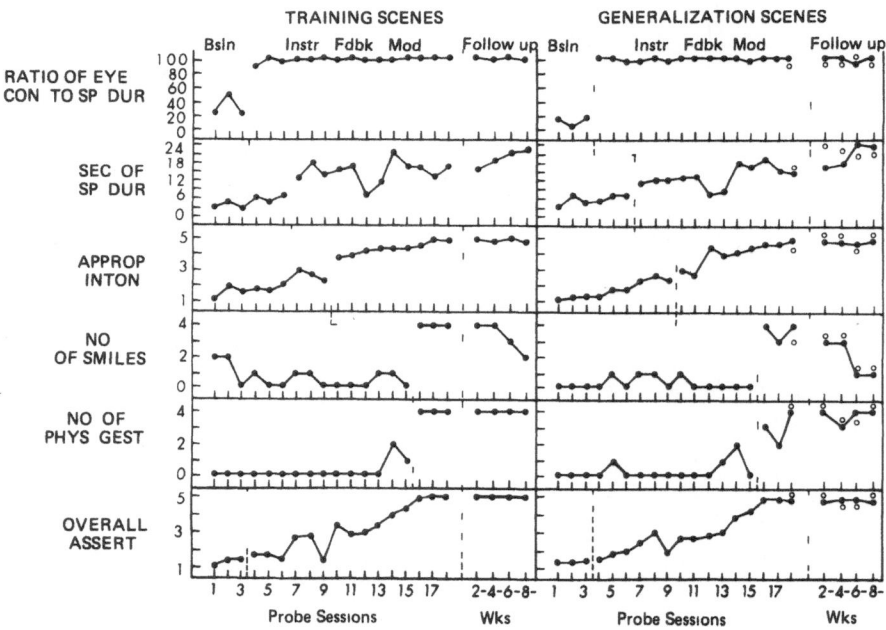

Figure 4. Probe sessions during baseline, treatment, and follow-up for Subject 1
From Bellack, Hersen, and Turner (1976), Figure 1

ments in targeted behaviors (when treatment was applied individually to each) were noted for probe measures of training scenes in addition to a linear increase in overall assertiveness. The data pattern is quite similar for the generalization scenes where no training was given. In addition, there appeared to be transfer of training to the novel scenes (represented by open circles on the graph). As for durability of results, follow-up data at 2, 4, 6, and 8 weeks indicate maintenance of improvement with the exception of some decrement in number of smiles. In summary, the specific effects of social skills training were positive, as were the generalization effects. (It might be noted that Bellack *et al.*, 1976 and Hersen and Bellack, 1976a have replicated their results with additional cases.)

An interesting application of the multiple baseline design procedure appeared in a paper by Liberman and Smith (1972). The patient was a 28-year-old married female who was attending the Oxnard Community Mental Health Center Day Treatment Program. She was suffering from multiple phobias that were quite incapacitating (i.e., fear of being alone, fear of menstruation and vaginismus associated with attempted insertion of tampons, inability to chew hard foods, and an extreme fear of having needed dental work performed).

Throughout baseline and treatment phases the patient participated in all scheduled activities of the day hospital program. During the baseline assessment she rated each of her fears on a Target Complaint Scale. Then, under a time-lagged basis, each of the phobias was treated with systematic desensitization (imaginal and *in vivo*).

The results of this single-case analysis are presented in Figure 5. The data clearly indicate the independence of the baselines, with improvement occurring only when treatment was specifically directed to each targeted fear. At the 2-month follow-up each of the fears had been eliminated with the exception of the fear of having dental surgery. However, there was some improvement in this fear to the point where the patient had actually been able to visit a dentist for a preliminary appointment. Moreover, general social functioning revealed marked improvement as well.

Future Directions

Analogous to the oft-posed question: "Does psychotherapy work?" it is equally fruitless at this time to ask: "Does partial hospitalization programming work?" In essence, this is the wrong question for researchers concerned with partial hospitalization to be asking. What, then, is the right question to be posed by such researchers? The answer

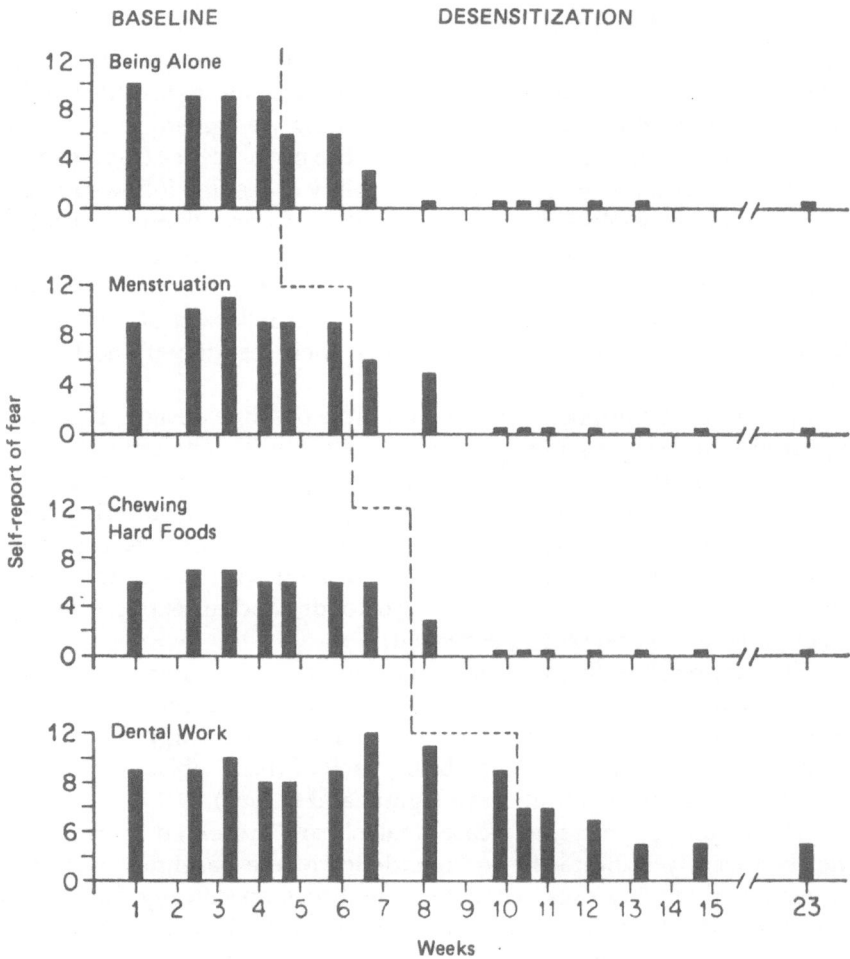

Figure 5. Multiple-baseline evaluation of desensitization in a single case with four phobias. From Liberman and Smith (1972), Figure 1.

was precisely articulated by Gordon Paul (1967) one decade ago in his classic paper, "Strategies of outcome research in psychotherapy." There he argued: "In all its complexity, the question towards which all outcome research should ultimately be directed is the following: *What* treatment, by *whom*, is most effective for *this* individual with *that* specific problem, and under *which* set of circumstances?" (Paul, 1967, p. 111).

With the exception of the few studies conducted by single-case researchers, Paul's (1967) recommended strategy for carrying out clinical research has not been followed by most partial hospitalization investigators. To the contrary, what little controlled research exists in this area has been done in most global fashion. Thus questions that follow have been neither posed nor answered: (1) What are the specific programmatic features in day hospital that maximize community adjustment for schizophrenics as opposed to depressives as opposed to character disorders attending day hospital? (2) What are the *effective* and *active* ingredients in a given partial hospitalization program? (i.e., Is it chemotherapy? Is it recreation therapy? Is it contingency management? Is it the therapeutic milieu in itself? Is it vocational guidance and training? Is it the therapeutic relationship? Is it social skills training? Or, is it some specific combination of the aforementioned for a given diagnostic group?) (3) Do goal attainment scaling (Austin *et al.*, 1976) and the problem-oriented approach to record keeping (Luber & Hersen, 1976) improve the diagnostic process, hence leading to more gratifying therapeutic results? (4) Are specific behavioral approaches to treatment more effective than "milieu" or recreational activity treatments? (5) What is the interface between chemotherapy and specific psychotherapeutic approaches for given diagnostic groups?

The above list represents just a few of the important questions that need to be posed, researched, and answered if the initial clinical impetus of partial hospitalization programming is to continue and become a more empirically documented treatment afforded to both acutely disturbed and chronic psychiatric patients. The present author sincerely trusts that many of these issues will be tackled by enterprising investigators in the next decade. If not, given the increasing empirical spirit pervading the psychiatric arena in general, partial hospitalization programming will undoubtedly suffer the fate of most undocumented therapeutic approaches (i.e., disfavor and decreased funding). This would be most unfortunate as partial hospitalization services have much to offer from both a rehabilitative and a preventive framework. However, it behooves the proponents of the concept to document this more fully.

Summary

This chapter has dealt with contemporary research issues facing the proponents of partial hospitalization. Empirical studies have been examined in a critical manner and were categorized under five separate headings: (1) patterns of utilization, (2) program evaluation, (3) comparison of partial hospitalization with inpatient hospitalization, (4) a comparison of behavioral and eclectic procedures, and (5) the single-case approach to research. In addition, future directions for partial hospitalization researchers were outlined in terms of some questions that remain unanswered in the field and many that have never been posed at all.

134   Michel Hersen

References

Alcohol, Drug Abuse, and Mental Health Administration *State and Regional Data, Federally Funded Community Mental Health Centers* Survey and Reports Branch, Division of Biometry, National Institute of Mental Health, Bethesda, Maryland, June 1974

Austin, N K , Liberman, R P , King, L W , & DeRisi, W J A comparative evaluation of two day hospitals *Journal of Nervous and Mental Disease*, 1976, 163, 253–262

Barlow, D H , & Hersen, M Single-case experimental designs Uses in applied clinical research *Archives of General Psychiatry*, 1973, 29, 319–325

Beigel, A , & Feder, S L Patterns of utilization in partial hospitalization *American Journal of Psychiatry*, 1970, 126, 101–108

Bellack, A S , Hersen, M , & Turner, S M Generalization effects of social skills training in chronic schizophrenics An experimental analysis *Behaviour Research and Therapy*, 1976, 14, 391–398

Cameron, D E The day hospital An experimental form of hospitalization for psychiatric patients *Modern Hospital*, 1947, 69, 60–62

Chasin, R M Special clinical problems in day hospitalization *American Journal of Psychiatry*, 1967, 123, 779–785

Craft, M Psychiatric day hospitals *American Journal of Psychiatry*, 1959, 116, 251–254

Eckman, T A *The educational workshop model as a treatment alternative in a partial hospitalization program* Unpublished manuscript, 1976

Eisler, R M , Hersen, M , Miller, P M , & Blanchard, E B Situational determinants of assertive behaviors *Journal of Consulting and Clinical Psychology*, 1975, 43, 330–340

Glasscote, R , Kraft, A M , Glassman, S , & Jepson, W *Partial hospitalization for the mentally ill A study of programs and problems* Washington, D C Garamond/Pridemark Press, 1969

Hersen, M , & Barlow, D H *Single case experimental designs Strategies for studying behavior change* New York Pergamon Press, 1976

Hersen, M , & Bellack, A S A multiple-baseline analysis of social-skills training in chronic schizophrenics *Journal of Applied Behavior Analysis*, 1976, 9, 239–245 (a)

Hersen, M , & Bellack, A S Social skills training for chronic psychiatric patients Rationale, research findings, and future directions *Comprehensive Psychiatry*, 1976, 17, 559–580 (b)

Hersen, M , & Bellack, A S Assessment of social skills In A R Ciminero, K S Calhoun, & H E Adams (Eds), *Handbook for behavioral assessment* New York Wiley, 1977

Hersen, M , & Luber, R F Use of group psychotherapy in a partial hospitalization service The remediation of basic skill deficits *International Journal of Group Psychotherapy*, 1977, 27, 361–376

Herz, M I , Endicott, J , Spitzer, R L , & Mesnikoff, A Day versus inpatient hospitalization A controlled study *American Journal of Psychiatry*, 1971, 127, 107–118

Hogarty, G E , Dennis, H , Guy, W , & Gross, G M "Who goes there?" — A critical evaluation of admissions to a psychiatric day hospital *American Journal of Psychiatry*, 1968, 124 94–104

Johnson, C A , & Kish, G B *A progress report of a longitudinal evaluation of the effectiveness of a partial hospitalization program* Paper presented at the Partial Hospitalization Study Group, Atlanta, Georgia, September 19, 1976

Lamb, H R Chronic psychiatric patients in the day hospital *Archives of General Psychiatry*, 1967, 17, 615–621

Liberman, R P , & Smith, V A multiple baseline study of systematic desensitization in a patient with multiple phobias *Behavior Therapy*, 1972, 3, 597–603

Lombardo, T. W., & Turner, S. M. Use of thought-stopping to control obsessive ruminations in a chronic schizophrenic patient. *Behavior Modification*, in press.

Luber, R. F., & Hersen, M. A systematic behavioral approach to partial hospitalization programming: Implications and applications. *Corrective and Social Psychiatry and Journal of Behavior Technology Methods and Therapy*, 1976, *4*, 33–37.

Michaux, M. H., Chelst, M. R., Foster, S. A., Pruim, R. J., & Dasinger, E. M. Postrelease adjustment of day and full-time psychiatric patients. *Archives of General Psychiatry*, 1973, *29*, 647–651.

Paul, G. L. Strategy of outcome research in psychotherapy. *Journal of Consulting Psychology*, 1967, *2*, 109–118.

Ruiz, P., & Saiger, G. Partial hospitalization within an urban slum. *American Journal of Psychiatry*, 1972, *129*, 121–123.

Salzman, C., Strauss, M. E., Engle, R. P., & Kamins, L. Overnight "guesting" of day hospital patients. *Comprehensive Psychiatry*, 1969, *10*, 369–375.

Sanders, S. H., Williamson, D., Akey, R., & Hollis, P. Advancement to independent living: A model behavioral program for the intermediate care of adults with behavioral and emotional problems. *Journal of Community Psychology*, 1976, *4*, 275–282.

Taube, C. A. *Day care services in federally funded community mental health centers 1971–1972*. Statistical Note no. 96, Survey and Reports Section, Biometry Branch, National Institute of Mental Health, Rockville, Maryland, October 1973.

Turner, S. M., Hersen, M., & Bellack, A. S. Effects of social disruption, stimulus interference and aversive conditioning on auditory hallucinations. *Behavior Modification*, 1977, *1*, 249–258.

Washburn, S. L., Vannicelli, M., Longabaugh, R., & Scheff, B. J. A controlled comparison of psychiatric day treatment and inpatient hospitalization. *Journal of Consulting and Clinical Psychology*, 1976, *44*, 665–675.

Wilder, J. F., Levin, G., & Zwerling, I. A two-year follow-up evaluation of acute psychotic patients treated in a day hospital. *American Journal of Psychiatry*, 1966, *122*, 1095–1101.

Wolpe, J. *The practice of behavior therapy* (2nd ed.). New York: Pergamon Press, 1973.

Zwerling, I., & Wilder, J. F. An evaluation of the applicability of the day hospital in treatment of acutely disturbed patients. *Israel Annals of Psychiatry and Related Disciplines*, 1964, *2*, 162–185

IV

Problems and Future Directions

Introduction

Over the course of its existence, several significant problem areas have predominated in the conceptualization and implementation of partial hospitalization. These problems have received some attention in the literature and have been briefly outlined in Chapter 1 of this book as well as in various contexts throughout other chapters. In this section the authors present a thorough analysis of these issues in order to clarify and evaluate their main features. In addition, suggestions for appropriate modes of action are made and steps toward the implementation of these suggestions are considered.

First, Samuel Turner discusses various administrative aspects of partial hospitalization programming; issues related to staffing patterns and the utilization of personnel representing various disciplines and varying degrees of expertise are explored; in addition the distinction between "day care" and "day treatment" is discussed and the utilization of expanded programs (i.e., weekend and evening components) is considered. Next, Paul Lefkovitz reviews the interrelationship between partial hospitalization programming and the patient population; issues related to heterogeneous and homogeneous populations, specialized patient populations, and the involvement of families in partial hospitalization treatment are discussed and their influence on treatment programming is explored. Third, F. Dee Goldberg and Joan Perrault discuss the pragmatic but crucial issue of funding for partial hospitalization services; background regarding the system of financing medical care in the United States is presented to provide a better understanding of the present status of partial hospitalization funding; important barriers faced by partial hospitalization programs in improving and expanding fiscal resources are presented and efforts that have successfully overcome some of these barriers are described. Finally, Raymond Luber presents an overview of the current status of partial hospitalization and explores some of the significant issues related to the future development and utilization of the treatment modality.

7

Treatment Orientation and Program Implications

Samuel M. Turner

Introduction

The advent of the first day hospital treatment program in North America at the Allen Memorial Hospital in Montreal (Cameron, 1947) marked the beginning of a new era in the treatment of psychiatric patients and in the way psychiatric illness is viewed. Prior to this innovation, the majority of psychiatric patients were treated on inpatient units at private and general hospitals or in state facilities that were usually placed in isolated areas away from the general population. When socially abhorrent symptoms were reduced (e.g., suicidal ideation, hallucinations, delusions) the patient was returned to his family and the community, where he was then treated on an outpatient basis, usually no more than once per week. Partial hospitalization programs or day hospitals proposed that patients be treated on a partial or day basis, eliminating the need for 24-hour care and keeping the patient in the community. Several years following the opening of Cameron's unit, day hospitals began to spring up in other areas of North America including the United States. Eventually evening programs were also made available for those individuals who were functioning well enough to participate in various activities (e.g., vocational training programs, attending school, working, actively job seeking, household responsibilities) but who were nevertheless in

Samuel M. Turner • Department of Psychiatry, Western Psychiatric Institute and Clinic, University of Pittsburgh School of Medicine, Pittsburgh, Pennsylvania 15261.

need of more intense treatment than is usually provided in outpatient therapy. The development of partial hospitalization programs was given added impetus with the passage of the Mental Retardation Facilities and Community Mental Health Centers Construction Act of 1963. The passing of this act more than any other single event signaled a shift away from institutional care to community-based treatment facilities. Moreover, the act required each community mental health facility to establish a day treatment center.

Proponents of partial hospitalization have long argued that the day treatment center provides many advantages over traditional institutional care. The most frequently cited advantage is the avoidance of the development of secondary institutional behaviors such as those described by Erwin Goffman in his now classic work, *Asylums* (Goffman, 1961). In addition, other benefits frequently cited are the use of the therapeutic community, family therapy, various activity groups, vocational counseling, intensive individual therapies, and various other group therapies, all while the patient remains in his usual social unit. Unfortunately, there has been little research to date to document that these assertions are, in fact, correct.

Patient Populations

Although the concept of day hospital is firmly entrenched in psychiatric treatment and their numbers are rapidly growing, there is still dispute as to what type of patient is most suitable for treatment in day centers. Are such treatment centers only for neurotic patients, or are more serious psychiatric disorders also amenable to treatment in such a facility? Are these centers suitable only for the chronic debilitated schizophrenic patients who have long histories of institutionalized care, or are they suitable for the treatment of acute disturbances when florid symptomatology is likely present? These questions as well as others continue to be the center of debate in terms of appropriate clientele for partial hospitalization programs. Several studies have been directed at examining the types of patients treated in day settings which may serve to shed some light on this problem. Hogarty, Dennis, Guy, and Gross (1968) studied 146 patients who were admitted to the Baltimore Psychiatric Day Center (BPDC). A variety of measurement instruments were used (Mental Status Schedule, Brief Psychiatric Rating Scales, Springfield Symptom Index, Kratz Adjustment Scales, Hopkins Distress Index). According to Hogarty *et al.*, the BPDC provided care for a very restricted range of patients who were white married females with high school educations, and living with a spouse who was a skilled or semi-skilled employee. Most of the patients (80%) were found to be of social class IV or higher with respect to the Hollingshead two-factor index of social class. Moreover, of the 85% with a previous history of psychiatric treatment, 50% received psychotherapy alone or in combination with other treatments; 46% of these patients had previous hospitalizations while 61% had only one hospitalization, usually of a brief duration (less than a month). These characteristics, of course, are representative of middle-class patients with relatively minor and nonpsychotic disorders.

When these 146 patients were compared with a sample of inpatients and outpatients, the following picture emerged. With respect to degree of psychopathology, BPDC patients were not similar to patients usually found on an inpatient service, but they were found to exhibit more pathology than was typically seen in patients frequenting outpatient clinics. Hence, they fell somewhere in the middle of the continuum of pathology ranging between severe and mild. Hogarty et al. concluded that patients treated in the BPDC were not the same as the typical inpatient in a psychiatric hospital. They further concluded that the day hospital tends to treat a select group of patients; thus it is not an alternative to inpatient treatment. While we have no reason to question data generated in this particular study concerning patients treated at the BPDC, we do question generalization based on these data about day hospitals in general. The authors have made the serious error of generalizing from a restricted sample (i.e., one center). That these data are not representative of all day programs is supported by several investigations briefly described below.

Odenheimer (1964) reported data on 100 patients who were admitted to the San Mateo County Health Services and subsequently screened for admission to the day treatment program. Gross contraindication such as lack of transportation, patient refusal to come to day hospital, and "inappropriate" family situation eliminated a large number and several others were eliminated due to unspecified reasons following an interview with the psychiatrist. Twenty-three of the 100 original patients were admitted to day hospital with presenting symptoms of delusions, hallucinations, alcoholism, suicidal behavior, depression, and other more bizarre symptomatology. Although there were no socioeconomic data available, all patients were judged to be appropriate for inpatient care during admission interviews.

Zwerling and Wilder (1962) reported that out of 72 patients requiring immediate psychiatric care, 23.6% were not accepted for day treatment due to lack of transportation and uncooperative families. Approximately 23.6% of the remainder eventually required treatment for various durations on the inpatient service. Six percent of these returned to the day hospital and completed treatment there, while 52.8% of all these patients were treated entirely in the day hospital and discharged to the community.

McMillan and Aase (1964) provided diagnostic and demographic data on the first 500 patients treated at the San Diego Day Treatment Center (SDDTC). Sixty-five percent of these patients were psychotic, 18% were diagnosed psychoneurotic, and 17% were diagnosed as personality disorders. Schizophrenic reactions of all types made up approximately 50% of all admissions. Using a stratification procedure, patients were assigned to one of six social class levels ranging from 1 = highest (professional managerial) to 6 = lowest (laborer) or to a seventh class (unknown). Twelve percent fell in class 1, 31% in class 2, 28% in class 3, 19% in class 4, 5% in class 5, 3% in class 6, and 1% unknown. Although it is difficult to determine just what these individuals' socioeconomic status was from the data presented, the authors suggested that the largest percentage of patients (class 2, 3, 4) represented roughly several ranges of the middle class.

From the data summarized above, it seems clear that patients of various diagnostic categories and diverse socioeconomic status are amenable to treatment in the day hospital setting. It cannot, on the other hand, be concluded from the data presented that day treatment centers are capable of totally replacing the traditional inpatient unit. To the contrary, though many of the patients reported in the above studies were treated on a partial basis, many required periods of inpatient hospitalization ranging from only an overnight stay (guesting) to several days or weeks. Salzman, Strauss, Engle, and Kamins (1969) examined variables related to overnight stays (guesting) of 137 day hospital patients at the Massachusetts Mental Health Center. Twenty percent of the patients required guesting during their course of treatment. The majority (68%) of these patients were diagnosed as schizophrenic and depressed, with acute schizophrenia representing the largest single group. In terms of reason for hospitalization, suicide precaution was the most frequent indication for guesting (57.1%). A host of demographic variables were examined in relation to guesting but none was significant at the usually accepted level of statistical significance (.05), except with respect to "difficulty disengaging from families."

It would appear that many patients who have in the past been treated on inpatient units can indeed be treated on a partial basis. However, inpatient units are still required when patients are deemed to be significantly suicidal, homicidal, or displaying other bizarre behavior that is unacceptable to families and/or the community. Other diagnostic categories such as organically impaired individuals and sexual deviates (e.g., rapists) might require 24-hour care for the benefit of themselves as well as society.

Day Treatment versus Day Care

It has been approximately three decades since the first day hospital program opened on the North American continent. Since the opening of that limited day treatment program in Montreal, many others have opened in the United States. Not all centers have been designed to treat the same patient populations, nor has treatment orientation or organizational structure been uniform. Many centers have specialized in the treatment of patients with chronic disorders (primarily schizophrenic), acutely disturbed patients, the mentally retarded, children with severe emotional disturbances, geriatric patients, and drug abusers. Just as day treatment centers have varied with respect to patient population, they have also differed with respect to location. Day centers have operated on the grounds of state mental institutions, in comprehensive community health centers, as part of psychiatric treatment facilities in general hospitals, in community mental health centers, and as completely independent entities. Likewise, in terms of administration, they have been operated as units of other programs (including inpatient units) as well as completely independent. Despite the widespread existence of such centers as well as a professional organization dedicated to the advancement of partial treatment (Federation of Partial Hospitalization Study Groups), there is as yet no clear perception within or without the mental health profession as to where this treatment modality fits into the overall mental health care system, what treatment modalities are most appropriate for what patient populations, how such programs should be staffed, or where they should be located. These problems, of course, have contributed to confusion in referral practices, staffing, and perhaps most importantly, in funding by third-party payers, be they state, federal, or private. Problems in funding for partial programs are discussed fully in Chapter 9 of this book. One of the primary problems plaguing partial programs has been the lack of clarity in distinguishing day hospital and day care centers. While some undoubtedly view the two as being synonymous, this need not be the case. The term *hospital* would seem to imply that active treatments of various types are available. On the other hand, day care seems to imply the existence of a program designed to provide a certain population with a place to go for certain hours during the day to socialize and engage in various other activities designed primarily to keep them busy. While it can be argued that both of the programs described above are desirable, problems in funding and referral practices exist due to the lack of clarity with respect to how the labels are used, and exactly what is involved with respect to treatment in each program. This is crucial since there is an obvious need for more specialized personnel and equipment in active day treatment centers.

It is clear, then, that there are important issues that must be resolved before the optimism generated by the day hospital movement
can be brought to fruition. Since it is unlikely that one program or one
treatment unit could adequately provide care for all of the various
types of patients in need of services, it seems reasonable that programs
should be structured with treatment of a particular type of patient
in mind.

Organizational Structure and Treatment Orientation

In 1951 Joshua Bierer outlined what he considered to be the optimal
organizational pattern for day hospitals. According to Bierer, a center
where chronic schizophrenic patients come for the day to be occupied
and to remain for extended periods of time should be called not a day
hospital but an occupational center for chronic patients. This type of
center would appear to be similar to day care centers, which were discussed earlier in this chapter. Aftercare rehabilitation centers should not
be called day hospitals, nor should day units in mental hospitals. The
latter, according to Bierer, should be called day wards. Facilities located
in general hospitals or teaching hospitals should be called day departments. Only facilities meeting the following criteria should be referred to
as day hospitals: "1. is an independent unit and a hospital in its own
right; 2. admit every type of psychiatric patient; 3. therefore, replace a
mental hospital; 4. provides all the treatments available in modern
psychiatry; 5. is built on the fundamental principles of the therapeutic
community, which means activation of the patient's abilities, therapeutic atmosphere, cooperation between patient and staff, and 'guided self
government'; 6. has therapeutic social clubs, each of whose weekly
meetings is regularly attended by one of the hospital's staff psychiatrists;
7. is patient-centered, and not dogma-centered" (Bierer, 1951). Obviously this description is not in line with current attitudes toward day
hospitals. Fortunately, chronic schizophrenic patients need not go to
centers that serve a "baby-sitting" function during certain hours of the
day. As we will see later, many day hospital centers have active
therapeutic programs for chronic schizophrenics as well as other patients with long histories of regressed behavior and maladaptive reper-
toires of various skills. Similarly, most mental health professionals
would agree that a day hospital program need not admit every type of
psychiatric patient or replace a mental hospital in order to be legitimately
referred to as a day hospital. Nevertheless, Bierer has made an attempt
to rigidly define what a day hospital should be with respect to organization, patients, and location. Unfortunately, this attempt has not been

followed or expanded, resulting in the current blurred status of day treatment programs. Following is a schema of day hospital programs with respect to appropriate patient populations, administration, staffing patterns, treatment modalities, and location.

Day Hospital for Acutely Disturbed Patients

Partial programs designed to treat the acutely disturbed psychiatric patient are likely to resemble the traditional inpatient unit more than any other type of day treatment center. Such programs must be prepared to deal with patients who manifest suicidal and homicidal ideation, as well as other bizarre and socially deviant behavior. Therefore, these programs must be staffed with large numbers of professional personnel, particularly medical personnel. Staffing should include psychiatrists, psychologists, social workers, and nurses in appropriate numbers to provide medical supervision and intensive therapy of various modalities. All forms of treatment should be available including pharmacological and somatic therapies. In this facility, the availability of some beds might be appropriate for limited use (e.g., guesting) as well as some rooms that could be used to isolate patients who might require brief separation from others. Therapeutic activities in this type of day hospital should be provided throughout the hospital period (just as it should in other types) based on the needs of the patient. Such therapy might include individual, group, family, marital, and recreational. Although there is an obvious need for a high degree of medical involvement (physicians, nurses) in this type of day program, the administrator, in my opinion, need not be a physician. On the other hand, he should be a good administrator, clinician, and researcher. While some might question the need for the leader of such a facility to be a researcher, with the current status of the science of mental health in general, and day hospital in particular, I believe it is essential that the leadership possess this quality. In order for a strong commitment to research to exist, it must emanate from the top.

The aim of programs for acutely disturbed patients is to replace or minimize use of the typical inpatient unit. However, as research has shown, many patients will require periods of inpatient treatment. Therefore, it seems logical that such programs might be physically attached to, or be in close proximity to inpatient facilities so that transfers can be accomplished rapidly and with ease. Patients should ideally spend about the same time in this program as they would in the typical acute inpatient unit. Following discharge, they would continue in outpatient therapy.

Day Hospital for the Chronic Patient

Most patients currently treated in day hospitals for the chronically ill carry the diagnosis of schizophrenia, many with histories of long-term continuous care in state hospitals. However, treatment in such facilities has also provided for a small percentage of patients with debilitating neuroses and character disorders. According to Astrachan, Flynn, Geller, and Harvey (1970), the primary aim of such a program is rehabilitation and reeducation. Medical consultation is essential but the facility need not be organized as a hospital unit with a preponderance of highly skilled nursing personnel. Indeed, what is needed are highly skilled therapists and educators with the ability to remediate interpersonal skill deficits and to provide specific instruction such as vocational counseling. Although individual marital and family therapy is included in the treatment, the primary focus of treatment is to rehabilitate and reeducate in every respect the debilitated chronic patient who is severely lacking in interpersonal, social, cultural, self-care, daily living, vocational, and recreational skills.

Eckman (1978) describes a day hospital program utilizing an educational workshop model at the Oxnard Mental Health Center in California. The program is described as an educational-behavioral program where a token economy system is used to motivate patients to participate in the various therapeutic activities. Patients participate in a variety of structured workshops including conversational skills, personal finance, consumerism, grooming, current events, ethnic exchange, public agencies, vocational preparedness, anxiety and depression management, weight control, and recreation-education-social-transportation. These groups or "workshops," as they are called, are designed to remediate deficits in specific areas and to facilitate integrating the patient into the mainstream of society. Similarly, the partial hospitalization program at Western Psychiatric Institute and Clinic (Hersen & Luber, 1977; Luber & Hersen, 1976) is designed to treat primarily the chronic patient by remediating basic interpersonal skill deficits. This program is behaviorally structured with a token economy program designed to motivate patients to attend various structured group activities. In this particular program, there is a strong emphasis on remediating basic interpersonal skill deficits through the use of social skills training. There are a host of other structured groups including grooming, current issues, theme-centered, bureaucratic systems, life management, and anxiety and depression management.

Clearly, the two programs briefly described above are designed to teach patients those basic skills that are needed to exist outside of the hospital setting. In my opinion, all programs designed to treat the chronic patient should be organized similarly. This particular type of program need not be attached or located close to an inpatient unit. As in the case of the acute program, leadership should be in the hands of a competent administrator, researcher, and clinician regardless of discipline.

Other Day Hospital Programs

A host of other specialized day hospital programs are also possible. These programs could be designed to meet the needs of very specific populations such as the mentally retarded, the physically handicapped, geriatric patients, and emotionally disturbed children. Indeed, there are such programs in existence in various places around the country. At this time, much of the day program for the mentally retarded should be similar to the one specified for chronic patients, featuring highly structured activities designed to teach specific skills. Programs for the physically handicapped should include physical therapy, muscle reeducation therapies (e.g., biofeedback), vocational counseling, interpersonal skills training, as well as individual, family, group, and marital therapies. Programs for the geriatric patient should include large recreational and social components. Such programs should also be equipped to retrain patients who have lost control of their sphincter muscles and are incontinent. Programs for emotionally disturbed children require skilled individual therapists, family therapists, medical personnel, and appropriate classroom settings. These programs have the dual role of providing therapeutic intervention and education simultaneously. There is no need for any of these programs to be located in or near other psychiatric treatment facilites. Nor is it particularly crucial as to the discipline of the leadership or administration. The primary criteria should be that he be experienced in the appropriate area with the attributes previously mentioned. Staffing patterns would include professionals from the various disciplines, various paraprofessionals, and in the case of centers for children and the mentally retarded, regular and special education teachers.

Cost of Treatment

Advocates of partial hospitalization have long invoked the cost factor in advocating partial treatment over full 24-hour inpatient treatment. It might appear that 24-hour inpatient treatment would cost more than treatment on a partial basis. However, partial programs in many instances tend to be more treatment-oriented (e.g., patients do not sit around all day as on many inpatient units), so that more skilled personnel are required as well as various ancillary personnel for transportation and other activities. Although lower cost is a theme heard constantly in literature on partial programs, surprisingly few comparative studies actually exist. Bierer (1962) provides some data on the cost of treatment for 2 years at the Marlborough Day Hospital in England. According to Bierer, complete psychiatric sevices in 1956–57 for Marlborough was $70,000 for 914 patients. In 1957–58, $87,000 was required to treat 1,207 patients. It is further reported that the cost of buying and equipping the buildings (2) was about $55,000, while the cost of building a mental hospital to treat 250 patients was estimated to be about $3 million in 1962. There were no direct comparative data offered, nor were specific types of provided treatment spelled out in detail. Goshen (1959) indicates that the per diem cost of treating patients in day hospital is higher than in standard state hospitals, but comparable to inpatient units in a general hospital. Furthermore, capital expenditure is considerably less than that required for building inpatient facilities because there is no need for sleeping facilities, housekeeping, feeding, etc. Again, no specific comparative data are offered.

In a more recent study, Washburn, Vannicelli, Longabaugh, and Scheff (1976) compared the cost for patients in a day hospital and an inpatient unit over an 18-month period. During the first 6 months, expenses for inpatient service exceeded day hospital cost considerably (mean = $11,525 and $8,670, respectively). During the second 6 months, inpatient cost averaged $7,122 while day hospital cost averaged $3,603. During the third 6-month period, inpatient cost averaged $3,269 while day hospital cost averaged $1,551. In this particular study, day hospital costs are consistently less than inpatient treatment. However, before definitive statements can be made, comparisons between specific types of programs providing specific types of services for specific types of patients must be made.

Summary

The initial excitement and optimism generated by the development of partial hospitalization as an alternative to traditional inpatient care has been tempered by the slow development of such programs and by confusion with respect to what constitutes a day hospital. Further confusion exists with respect to what patient populations are appropriate, where the centers should be located, what constitute appropriate treatment orientations, and how such programs should be administered and staffed.

Although research into the effectiveness of partial programs is scarce, it seems clear that many patients formerly treated on inpatient units can be treated on a partial basis. Furthermore, the cost of treatment is probably much less. However, partial programs cannot totally replace inpatient units, which are needed for patients who are extremely detrimental to themselves or others, or who because of other reasons cannot be seen on a partial basis.

There is a need for specific day hospital programs to be developed for specific patient populations. This would include programs for acutely disturbed patients, chronic patients, retarded patients, geriatric patients, emotionally disturbed children, physically handicapped individuals, and patients in need of aftercare services. Staffing patterns must be geared to the needs of the patients and the types of services offered rather than to a specific professional discipline. Determination of the effectiveness of such programs requires this type of specificity as well as an empirical attitude on the part of the administration and staff, particularly on the part of the director.

References

Astrachan, B M , Flynn, H R , Geller, J D , & Harvey, H H Systems approach to day hospitalization *Archives of General Psychiatry*, 1970, *22*, 550–559

Bierer, J The day hospital An experiment in social psychiatry and syntho-analytic psychotherapy London Washburn & Sons, 1951

Bierer, J The day hospital Therapy in a guided democracy *Mental Hospitals*, 1962, *3*, 246–252

Cameron, D E The day hospital *Modern Hospital*, 1947, *69*, 60–62

Eckman, T A Behavioral approaches to partial hospitalization In R Luber (Ed), *Partial hospitalization A current perspective* Kalamazoo, Michigan Behaviordelia, 1978

Goffman, E *Asylums* Garden City Doubleday, 1961

Goshen, C L New concepts of psychiatric care with special reference to the day hospital *American Journal of Psychiatry*, 1959, *115*, 808–811

Hersen, M , & Luber, R F Use of group psychotherapy in a partial hospitalization service The remediation of basic skills deficits *International Journal of Group Psychotherapy*, 1977, *27*, 361–376

Hogarty, G E , Dennis, H , Guy, W , & Gross, G M "Who goes there?" — A critical evaluation of admissions to a psychiatric day hospital *American Journal of Psychiatry*, 1968, *124*, 94–104

Luber, R F , & Hersen, M A systematic behavioral approach to partial hospitalization programming Implication and applications *Journal of Corrective and Social Psychiatry and Journal of Behavior Technology Methods and Therapy*, 1976, *22*, 33–37

McMillan, T M , & Aase, B H Analysis of first 500 patients at San Diego Day Treatment Center In R Epps & L Hanes (Eds), *Day care of psychiatric patients* Springfield, Illinois Charles C Thomas, 1964

Odenheimer, J. F. Day hospital as an alternative to the psychiatric ward. *Archives of General Psychiatry*, 1964, *13*, 46–53.

Salzman, C., Strauss, M. E., Engle, R. P., Jr., & Kamins, L. Overnight "guesting" of day hospital patients. *Comprehensive Psychiatry*. 1969, *10*, 369–375.

Washburn, S. L. Vannicelli, M., Longabaugh, R., & Scheff, B. J. A controlled comparison of psychiatric day treatment and inpatient hospitalization. *Journal of Consulting and Clinical Psychology*, 1976, *44*, 665–675.

Zwerling, I., & Wilder, J. F. Day hospital treatment of psychotic patients. In J. H. Masserman (Ed.), *Current psychiatric therapies* (Vol. 2). New York: Grune & Stratton, 1962, pp. 200–210.

8

Patient Population and Treatment Programming

Paul M. Lefkovitz

Introduction

The "personality" of a partial hospitalization service is shaped and influenced by a myriad of factors. However, one element in particular that is crucial to the understanding of partial hospitalization programming is the patient population. The collective needs of the patients, clients, or members, as they are variously called, ultimately determine the trends that will occur in designing and implementing the clinical services. This chapter will explore the interrelationships between the patient population and partial hospitalization programming from a variety of vantage points.

The issue of primacy will initially be considered; should the partial hospitalization program be designed to meet the identified needs of a given patient population or should the nature of the patient population be determined by the type of day program that is desired? In some instances, the program itself is planned first, being built around a particular therapeutic philosophy, staffing pattern, or administrative/budgetary consideration. The patient population that subsequently emerges will tend to reflect those formative factors. In other instances, the program is planned after assessing the particular clinical needs of the

Paul M. Lefkovitz • Gallahue Mental Health Center, Community Hospital of Indianapolis, Indianapolis, Indiana 46219.

catchment area or treatment facility. As a result, the characteristics of the program will reflect the needs of the identified patient population.

There appears to be no clear-cut superiority of one logistic approach over the other. Certainly, the two approaches may receive equal and concurrent consideration as they are not mutually exclusive. It seems that factors intrinsic to individual mental health settings should govern the overall approach in determining whether to first identify the target population or the type of day program to be designed. One observation that can be made, however, is that the option given the highest priority will have an inextricable impact on the other.

Homogeneous and Heterogeneous Patient Populations

Partial hospitalization's far-reaching applicability with a wide variety of patients has, ironically, given birth to one of the more spirited controversies in the field. Day programs have demonstrated an impressive degree of success with populations ranging from the most acute (Zwerling & Wilder, 1964) to the most chronic (Meltzoff & Blumenthal, 1966). Most clinicians would agree that the therapeutic needs of those two types of patient populations differ markedly (Gootnick, 1971). As a result, certain writers have suggested that there should be two different kinds of partial hospitalization programs, one for acute patients and one for chronic patients (Astrachan, Flynn, Geller, & Harvey, 1971; Beigel & Feder, 1970). Others, however, have maintained that the effective treatment of both types of patients can be carried out within the structure of a single day program (Lamb, 1967). This section will explore the advantages and disadvantages associated with both of those positions. Empirical data bearing upon this issue will be presented and recommendations will be made.

In using the terms *acute* and *chronic*, certain ambiguities regarding their exact connotations may arise. Without a doubt, there are no clearcut distinctions between the terms that would easily and successfully classify all patients. However, in most instances, it is reasonably understood what is meant when "acute" and "chronic" are used to describe patient populations. In a chronic population, the duration of the dysfunction appears to be one of the cardinal traits. Patients in that group generally display symptoms of a schizophrenic process that persist or reoccur over an extended period of time. That process is typically characterized by significant deficits in basic daily living skills as well as severe limitations in the area of interpersonal functioning. The clinical state of the patient at the time of treatment is also of major significance. When the current symptoms appear to be a reflection of an unchanging pattern of dysfunction with little recent variation, a chronic process is indicated. On the other hand, a disturbance marked by the *recent* development of major symptoms or the exacerbation of prior symptoms (even in an otherwise chronic profile) would be regarded by most as an acute disorder.

The acute disorder, therefore, is generally characterized by recent onset or exacerbation of symptoms. Frequently, although not necessarily, some specific precipitating event can be identified. Typically, there is a fair amount of diagnostic diversity in an acute population with a higher proportion of psychoneurotic disorders than in a chronic group. Acute psychoses of both the affective and cognitive variety would generally be represented as well. The overall functional level would also sometimes distinguish the acute and chronic groups. Although the behavior of the acute patient might be temporarily disorganized and severely impaired, there frequently is a history of at least marginal premorbid adjustment. When the most florid or intense symptoms of the acute phase subside, the majority of these patients will not present significant deficits in skills of daily living.

The terms *homogeneous* and *heterogeneous* will be used to define the two major types of patient populations. A heterogeneous population exists in a single day program that provides treatment to patients without regard to the chronicity or acuteness of their symptoms. In this type of program, there would be a very diverse variety of functional levels represented. A homogeneous population, in contrast, would be associated with a day program that specializes primarily in the treatment of *either* acute or chronic disorders, but not both. Accordingly, there would tend to be greater commonalities in the histories, functional levels, and, possibly, symptomatology within this patient population.

The Homogeneous Population

Several years ago, in the 1960s, the Veterans Administration began placing emphasis on two different types of day programs. One was referred to as the "day hospital." Its main goal was to provide short-term treatment to patients in acute crisis. The other type, referred to as the "day treatment center," was established to provide rehabilitative treatment to chronic patients (Ognyanov & Cowen, 1974). These two types of programs were designed to be administered and staffed separately in order to meet the specific needs of their respective patient populations. This action taken by the VA represents the broadest effort thus far to foster homogeneity in the patient populations of partial hospitalization programs.

Some empirical support for the concept of homogeneity in patient groupings is provided by Beigel and Feder (1970), who conducted an investigation of the patterns of utilization in a day program designed chiefly for acute patients. In this study, records were reviewed to identify patients who were admitted directly to the program from the community in a given year, excluding transfers from the inpatient service and private practice referrals. A resultant N of 176 was obtained. Subjects were then divided into two groups distinguished by complete or incomplete utilization of the program. Incomplete utilization was defined as transfer to the inpatient unit without subsequent return to the day center, withdrawal from the program by the patient, or removal due to disruptiveness. Of the 176 patients, 115 completed the program, while 61 did not. Several variables were examined to determine what factors distinguished complete utilization from incomplete utilization. Among them were suicide potential, family involvement, diagnosis, history of prior hospitalization, and clinical state of the patient. It was found that the clinical state of the patient, in terms of acuteness or chronicity, was the *only* factor significantly related to complete utilization of the program ($p < .001$). The sharp contrasts in the levels of incomplete utilization between chronic patients (58%) and acute patients (24%) led the investigators to conclude the following:

> This suggests that it is *very important* for a partial hospital to develop its treatment program and adjust its selection criteria according to well-defined goals. Two types of programs should evolve for the acute and chronic patients, and they should be quite different. (p. 1270)

The program envisioned for acute patients was described as an intensive treatment facility staffed by nurses, social workers, psychologists, and psychiatrists. It would be located near the inpatient service, although not necessarily in the same building. The primary treatment modes would be individual and group therapy.

In contrast, the center for chronic patients would emphasize a more supportive approach and take place over a more extended period of time. Its location would be in the community rather than near the hospital. The staff would consist of social workers, indigenous mental health workers, and activity therapists. The program would focus primarily on the development of daily living skills and social and vocational rehabilitation.

Further analysis of the data led to some speculation about the reasons for incomplete utilization among chronic patients. It was suggested that "the chronic patients saw the active treatment program as a threat to their long standing regressive needs" (p. 1271). The authors concluded that the availability of two kinds of day programs might broaden the range of patients who could benefit from partial hospitalization.

Astrachan *et al.* (1971) explored the issue of homogeneity from an entirely different point of view. Although their position lacks empirical support, it does represent a systematic and methodical investigation of the topic employing a systems approach. Systems theory was used to identify four major tasks of partial hospitalization programs: (1) providing crisis-oriented treatment as an alternative to hospitalization, (2) facilitating reentry into the community, (3) providing rehabilitative treatment for the chronic patient, and (4) delivering specialized services (i.e., geriatric, substance abuse) that are needed in the community.

The authors maintain that it is essential for a day program to clearly identify its *primary* task and that "all secondary tasks will . . . inhibit the ability of the day hospital to do its primary task most effectively" (p. 171). They explored the "goodness of fit" between the four primary tasks they identified and concluded, "It is extremely difficult to see how all of the tasks we have listed can be accommodated within any single day hospital structure" (p. 182). They did feel that the task of providing transitional services could be efficiently integrated with others. However, they seemed to fully support the notion of separate day programs for acute and chronic patients. In delineating two different types of day programs, they emphasized the same distinctions noted by Beigel and Feder (1970).

The program for chronic patients described by Astrachan *et al.* was rehabilitative in its approach with emphasis on socialization and vocational experiences. Nursing and occupational therapy skills were considered important in the staffing pattern. The location of the center was not thought to be significant. The program for acute disorders was described as being similar to a hospital in the kinds of services provided. The need for inpatient backup was stressed suggesting that its placement should be near the hospital. The staff in this type of program, it was felt, should be "clinically sophisticated" (p. 178). Therefore, these authors reached essentially the same conclusion as did Beigel and Feder in supporting homogeneous populations in day programs. It is significant that they arrived at these similar positions from very different and unique perspectives.

Both Astrachan *et al.* (1971) and Beigel and Feder (1970) seemed to base their recommendations on the disadvantages they perceived in treating a heterogeneous population. However, there are certain advantages to maintaining separate programs which they did not fully address.

Perhaps the most persuasive argument in favor of developing separate programs for chronic and acute patients is that such an approach allows for greater attention to the unique clinical needs of those two groups. The nature of the therapeutic program can then be "tailor-made" to the specifications of the population being treated. In an acute, crisis-oriented program, there can be greater emphasis upon verbally oriented therapies, affective arousal, in-depth insight, and interpersonal intimacy. In the chronic program, more attention can be given to skills of daily living, basic communications, vocational rehabilitation, recreational and occupational therapies, and supportive treatment. The advocates of separate programs have pointed out that it is difficult, at best, to provide a full range of needed services within a heterogeneous population. Therefore, the homogeneous setting may be more likely to provide therapeutic experiences that are relevant, meaningful, and at a manageable comfort level to the patients.

It also seems likely that cohesiveness and trust would be enhanced by the homogeneity of the population. Patients who are functioning at a similar level find it easier to relate to one another. Risk taking and self-disclosure are then facilitated within an atmosphere of mutual identification.

Designing a day program to treat a homogeneous group of patients is not without its disadvantages, however. The cost involved in developing and maintaining two separate day centers may be entirely prohibitive for many mental health settings. Also, in view of the frequent underutilization of partial hospitalization (Glasscote, Kraft, Glassman, & Jepson, 1969), the support in terms of referrals may be insufficient for the maintenance of two separate programs.

Difficulties may be encountered in staffing a homogeneous day program when the population being served is a chronic one. The limited treatment expectations appropriate for this group can become quite frustrating for the clinician. Klein (1974) addresses this issue in a most poignant manner.

In a system containing two separate day centers, considerable confusion could conceivably evolve in deciding which program is appropriate for a given patient. Regardless of how specific the guidelines are, there will always be a number of patients who elude a straightforward disposition. This would be especially true with patients presenting acute symptoms within an otherwise chronic profile. It is even possible that certain clinicians might refer some of those patients elsewhere — for example, to the hospital — rather than agonize over which day program to utilize.

In summarizing the advantages and disadvantages of the homogeneous approach to partial hospitalization programming, it appears that the major arguments in support emphasize clinical benefits while the basic arguments against revolve around financial and logistical difficulties. Additional considerations come to light, however, in exploring the advantages and disadvantages of treating a heterogeneous patient population.

The Heterogeneous Population

Empirical support for the concept of heterogeneity in the patient population is offered by Lamb (1967), who conducted a retrospective study of the "successful" completion of day treatment based upon the hospital records of 151 patients. The completion rate of "chronic patients" was compared with that of "nonchronically disabled" patients. It was found that the completion rates for chronic patients (64.7%) and "nonchronic" patients (68%) did not differ significantly. The author concluded that "the effective treatment of the chronic patient can proceed side by side with treatment of the acute patient in the same day hospital" (p. 616).

Lamb went on to assert that, in some respects, it is advantageous to integrate acute and chronic patients in the same setting. The value of modeling is a primary consideration. The opportunities for positive imitative behavior are greatly enhanced in a population represented by various functional levels. Chronic patients tend to use the higher functioning individuals as role models and thereby raise their level of social and vocational performance to those higher levels. This can be especially crucial for certain individuals who find it too threatening to attempt to identify with members of the staff.

Another major advantage in maintaining a heterogeneous population is that it facilitates flexibility and continuity of care. As patients make significant therapeutic progress and increase their overall level of functioning, they are less likely to "grow out of" a broadly based program treating a diverse population. There should always be some element of the program that is appropriate to their needs.

The obvious financial advantages of treating a heterogeneous group cannot be overlooked. The need for only one facility or physical location presents a substantial cost saving over two separate programs.

The major arguments in opposition to the heterogeneous population stem from clinical complications associated with the simultaneous treatment of chronic and acute patients. The rapid progress displayed by the acute patients can be highly discouraging and frustrating for chronic patients. In Lamb's (1967) study, the nonchronic group completed the program in one-half the time of the chronic group. The verbal ability, capacity for insight, and motivation of the acute patient can present a challenge so impossible and threatening to the chronic patient that he may simply give up or drop out of treatment. This factor could easily have contributed to Beigel and Feder's (1970) findings regarding the high dropout rates for chronic patients.

The phenomenological picture from the acute patient's point of view may be equally discouraging and frustrating. This type of patient might feel "held back" by the less appropriate involvement of the chronic patients in groups and other activities. In some instances, the acute patient might feel demeaned by his association with people functioning at a very regressed level. To an extent, it is true that the inappropriate behavior of the lower functioning patients may impede the progress of higher functioning individuals when they are in the same groups. Therefore, it might be difficult to achieve an overall balance between the clinical needs of acute and chronic patients when treating both populations simultaneously.

The feelings of being overwhelmed or held back, described above, sometimes motivate patients in a heterogeneous population to spontaneously form separate "cliques" that are roughly based upon chronicity or functional level. Although these cliques are only rarely openly hostile to one another (and, in fact, may not even be discernable to the staff), they can disrupt the overall cohesiveness of the therapeutic milieu. Therefore, a trusting and cohesive treatment environment may be somewhat more difficult to achieve with a heterogeneous population.

It appears, therefore, that the major problem to be considered with a heterogeneous group is the extreme difficulty in providing a program of therapeutic activities that is equally helpful to the entire range of patients. As previously noted, the therapeutic needs of acute and chronic patients frequently differ dramatically. Consequently, any single program of treatment is bound to lose relevancy for certain patients.

To deal with this most critical problem, some workers have developed the concept of "subprogramming" within the overall treatment framework. Washburn (1976) implemented such an arrangement and referred to it as a "track system." He emphasized that this allows patients of varying functional levels to be involved in therapeutic pursuits appropriate to their needs. He also pointed out that the system, being highly individualized, makes it possible to go beyond the basic acute/chronic distinctions in determining what would be most helpful for a given patient.

Other practitioners have developed day programs with separate "levels" designed to meet the specific needs of acute and chronic patients. For lack of any agreed-upon term to describe such a system, this approach will be referred to as "stratified" programming. Although individuals in stratified programs are considered part of the same overall milieu and participate in some activities together, there is a conscious effort to facilitate primary identification and cohesiveness *within* each "level." Separate programming is then developed and implemented for each subgroup.

Stratified Programming

Typically, the distinctions made in programming for the various levels within a stratified system reflect those suggested by Beigel and Feder (1970) and Astrachan *et al.* (1971). The level of programming designed for the higher functioning, acute patients generally emphasizes a relatively short-term, fast-paced, structured, probing approach with a good deal of group therapy involvement. Besides open-process groups, other activities offered within this level tend to be more verbally oriented, such as psychodrama, assertiveness training, communication skills groups, and goal-setting groups, among others. Specialized groups that are based on a particular ideology may be included, such as Gestalt groups, Rational-Emotive Therapy groups, and Transactional Analysis groups.

Programming for the level serving the chronic patients would emphasize a more supportive approach, although it might be just as structured. The patient would more frequently require the vehicle of task-oriented group activities to facilitate interaction. Therefore, recreational and occupational therapy would play a much more important role at this program level. The pace would be more deliberate and confrontive techniques would be used more sparingly. The average length of stay would typically be appreciably longer than in the more acute level. Rehabilitation would be a major focus and an important program component would revolve around teaching basic skills of daily living (shopping, cooking, hygiene, housing, etc.). Vocational guidance and assistance would also be essential. Fundamental interpersonal skills would typically be focused on throughout the milieu. A knowledge of relevant community resources (i.e., medical, legal, governmental, and recreational organizations) would generally be fostered through discussion groups and actual field trips.

In essence, developing a stratified program with differentiated "levels" represents somewhat of a compromise measure between the "pure" homogeneous and heterogeneous approaches. It successfully deals with the major difficulty of meeting the varying needs of a diverse clinical population within a single setting. Yet this approach retains the financial and logistic advantages of requiring only one day facility.

Stratified programming also retains some of the other advantages of the two basic approaches. For example, the opportunities for modeling, one of the basic strengths of the heterogeneous grouping, would still be present. Also, the benefits of increased flexibility and continuity of care would still exist because of the opportunities for movement between levels. Such movement can be facilitated by building in as many levels as are appropriate to the patient population. Cohesiveness, which can be significantly diluted in a heterogeneous program, should present fewer difficulties in a stratified system. Although identification with the overall milieu would be weakened, cohesiveness within each level can be very potent. This approach also assists individuals in working at a comfort level that is within reasonable limits. For clinicians making referrals, a stratified day program would be less confusing than two separate programs treating homogeneous groups. The type of programming required by a given patient could be determined by the day center staff after the referral is made rather than by the referring therapist (who might be less knowledgeable regarding such matters).

The most basic argument against a stratified system is, once again, a financial one. Additional staff are required to conduct the extra groups when two or more levels are functioning simultaneously. In some day programs, efforts have been made to provide stratified services in the absence of adequate manpower. One approach has been to eliminate cotherapy teams and, instead, to have staff conduct more sessions individually. Another has been to provide services to different levels at different times, e.g., morning versus afternoons, different days. Still another has been to stratify only a portion of the therapeutic activities, usually group therapy, while leaving the rest of the program unchanged.

These measures, especially the latter two, are generally not effective in bringing about the advantages of a true stratified system. Reducing the overall weekly therapeutic contact in order to achieve stratification would probably diminish the intensity of the program too greatly. Also, creating separate levels for only some therapeutic sessions would not provide enough contact to bring about sufficient cohesiveness within that subgrouping. It appears that approximately one-half of the treatment program should foster interaction *within* each level in order to facilitate adequate trust and cohesiveness.

Another disadvantage of the stratified system is the degree of effort that must go into program organization and planning. Maintaining two or more "subprograms" within the same facility by the same staff can become quite confusing and exhausting. Astrachan *et al.* (1971) pointed out that a system that pursues multiple tasks is likely to encounter staff conflicts as clinicians come to represent and identify with different aspects of the program.

In summary, a stratified approach to partial hospitalization programming can reduce some of the difficulties associated with a purely homogeneous or heterogeneous approach. However, it is not without its own disadvantages. The following section will present an overview of the three approaches to programming and provide some basic recommendations.

Homogeneous versus Heterogeneous Populations: An Overview

The empirical work that has explored homogeneity versus heterogeneity in patient populations is of limited utility in providing definitive guidelines to the partial hospitalization professional. The extreme paucity of research in this area is one limiting factor. Also, the available research that has been cited (Beigel & Feder, 1970; Lamb, 1967) has resulted in inconsistent and contradictory findings. To reiterate, Beigel and Feder found that in a heterogeneous population there were significant differences between acute and chronic patients in the rate of successful completion of the program. Lamb, in direct contrast, found no such differences. However, it may be possible to at least partially reconcile these contradictory results.

First of all, the nature of the day programs in which those two studies took place appears to have differed. Beigel and Feder described their center as an intensive, short-term treatment program "where the amelioration of acute symptoms was the major treatment goal" (p. 1270). Lamb's program was not comparably oriented. As Beigel and Feder themselves pointed out, it was hardly surprising to find that chronic patients tended not to complete a program that was primarily designed for the treatment of acute disorders. Therefore, the results of the two studies, carried out in such different settings, may not be directly comparable.

Also, Lamb's study seems to suffer from a lack of specificity in describing his patient population. His designation of "nonchronically disabled" (average duration of disability was 3.1 months) implies that a truly "acute" population was not rigorously identified and employed in the study. Therefore, a meaningful comparison between "acute" and "chronic" patients may not have been carried out. It appears that the findings of both of these studies may need to be qualified somewhat.

Lamb's findings may actually only suggest that chronic patients and nonchronic patients *within a certain range* can successfully be treated together in the same program. Beigel and Feder's findings indicated that chronic patients cannot be treated effectively in a short-term program designed for acute disorders. Therefore, it appears that the two studies addressed slightly different aspects of the issue and that their results may not be contradictory. Neither investigation actually compared the outcome of clearly defined acute and chronic populations under controlled, unbiased conditions. Consequently, the practitioner is left without firm empirical evidence in support of homogeneity or heterogeneity in patient populations. A well-controlled study is still needed to provide substantive data bearing upon this issue.

In spite of the limited assistance the available data provide in resolving this controversy, some general guidelines and recommendations can still be offered. In the "ideal" situation, where funding is ample and the treatment system is flexible, it appears that the homogeneous treatment population presents the best opportunity for creative and effective clinical programming. This inference is based on the significant number of advantages associated with the homogeneous approach as well as on the therapeutic disadvantages identified with the heterogeneous population. If budgetary considerations allow for the development of separate programs for acute and chronic patients, that would be the course of action this writer would recommend.

Of course, the real world is not always an "ideal" place, and professionals in the area of partial hospitalization have frequently been confronted with inadequate funding. In the absence of sufficient financial support, it is believed that the stratified approach would serve as the next most attractive option. Some of the most important assets identified with the homogeneous approach still exist within a conscientiously planned program of this nature. The additional staff required in a stratified program could possibly be composed of ranks of volunteers and professionals principally assigned to other areas of the mental health facility. Therefore, if funding does not permit the development of two separate day programs, a creatively pursued stratified approach can serve as an efficient alternative.

If financial support is so limited that a stratified program cannot be developed, the heterogeneous approach can still certainly bring about effective results. However, it may be helpful to keep in mind some of the difficulties and pitfalls encountered with this approach in order to anticipate and resolve them successfully. One way of reducing some of those difficulties might be to reduce the heterogeneity of the population. That is, it might be feasible to treat some of the most acute and chronic patients elsewhere (i.e., short-term hospitalization or outpatient aftercare). Such an approach may not necessarily be most helpful to the individuals at either extreme, but it could prove beneficial to the majority of the other patients.

It must be remembered that these recommendations are not supported by firm empirical data. They come primarily from the clinical experiences of the author and other professionals in the field. Therefore, the importance of conducting systematic inquiry into this question cannot be overemphasized. It is an unfortunate footnote to the field of partial hospitalization that such a fundamental question regarding the nature of the patient population has not been more thoroughly investigated. Although this issue is a difficult and complex one, it is hoped that the current discussion may stimulate such endeavors in the future.

Specialized Treatment Populations

An extension of the concept of homogeneity is the specialized treatment population. The advantages that are associated with treating a homogeneous population have encouraged the development of unique day programs for people with special needs. In some instances, these special needs have been defined by age, as in the implementation of day programs for children, adolescents (Laufer, Laffey, & Davidson, 1974), and geriatric groups (Farndale, 1961). Day programs have also been established for the treatment of specific difficulties or particular diagnoses. Harrington and Mayer-Gross (1959) describe a program specifically designed for neurotics in an industrial community, while MacKenzie and Pilling (1972), from the Mayo Clinic, report on a day program for out-of-town patients with neurotic and psychosomatic problems. Partial hospitalization has also been extended to nonpsychiatric populations with considerable success. Reports have appeared in the literature on the use of day programs for diabetics (Gordon & Weldon, 1973), for individuals with obesity (Westlake, Levitz, & Stunkard, 1974), and for those with epilepsy (Luber, Adebimpe, & Trocki, 1976). This section will explore the ways in which partial hospitalization has been utilized with these specialized populations. Also, a brief review of other groups that could benefit from the potency of this modality will be presented.

Programming for Specialized Psychiatric Populations

Harrington and Mayer-Gross (1959) reported on a specialized program to treat a basically neurotic population located in an industrial community. Although the inception of this program came rather early in the development of partial hospitalization, it represented a sophisticated attempt to meet the unique needs of their patient community. Group therapy, relaxation therapy, and other activities appeared to be directly geared to the requirements of their treatment population. Also, the fact that the program was located in an industrial community seemed to add to the emphasis on vocational rehabilitation. The authors reported that among 30 patients who were persistently unemployed before treatment, 15 entered gainful employment upon discharge from the day program.

The day program at the Mayo Clinic was designed chiefly for patients with neurotic and/or psychosomatic difficulties (MacKenzie & Pilling, 1972). Their programming had to be geared not only to the needs of a specialized clinical group but also to the fact that many of their referrals came from out of town. The result was a short-term, intensive program that focused on psychodynamically oriented group and individual therapy. The average length of stay was 2 to 3 weeks. The authors described their program as unique, although in actuality it closely resembles the short-term program previously described for acute patients (Beigel & Feder, 1970).

The programming described by MacKenzie and Pilling evolved through several stages. Initially, the schedule of therapeutic activities consisted of a daily community-oriented meeting, one small-group session per week, 2 hours of individual therapy per week, and one psychodrama session per week. Structured occupational therapy and lectures were to contribute to the rest of the treatment program. The staff found, however, that since most of the patients were well motivated to delve into their problems, "the initial enthusiasm for organizing and executing projects gave way to a philosophy that it was more useful for the patients to verbalize and discuss their personal and social difficulties than to engage in joint activities" (p. 358). Eventually, the program came to offer 2 hours of small group therapy sessions per day, 1 or 2 hours of occupational therapy per day, one large summary group meeting per day, psychodrama once per week, and an unspecified amount of individual psychotherapy. This program represents a good example of the tactic described at the beginning of this chapter in which a target population is identified first and then subsequent programming emerges from the perceived needs of that group.

Day centers established chiefly for the chronic schizophrenic have already been alluded to in this chapter (Gootnick, 1971; Weinstein, 1960). As previously mentioned, these specialized programs are able to provide "tailor-made" treatment to chronic patients on the basis of length of stay, staffing patterns, and types of psychotherapies employed.

Other authors have discussed the unique problems encountered in programming for certain psychiatric groups that are integrated as part of a broader patient population. Hyland (1977) presented an overview of the treatment of the borderline personality in a day program. In like manner, Emerson and Fagan (1977) explored the programming considerations that need to be taken into account in treating the mentally retarded. (See also Chapter 5 of this volume.)

Programming for Nonpsychiatric Populations

Certain types of disorders that are not "psychiatric" in the strict sense of the word have been treated effectively in day programs. Diabetes, obesity, epilepsy, and adjustment to old age all present difficulties that are very similar to those presented by psychiatric patients in many respects. With each disorder, there is usually limited insight into the nature of the problem, in terms of both etiology and its effects on daily living patterns. There is also frequently a feeling of helplessness and discouragement and of being out of control. The patient may also feel alone, misunderstood, or unappreciated. Some basic skills of daily living may have been altered or lost in the midst of the difficulties. The individual may lack adequate knowledge about how to cope with the unfortunate situation.

A partial hospitalization program designed for these patient populations, regardless of the etiology of the disorder, can provide the necessary insight, hope and encouragement, support, new coping mechanisms, and skills of daily living to enrich their lives and raise their level of involvement and functioning in the community.

In 1969, a major program for establishing diabetic "day-care clinics" in several hospitals in Nova Scotia was undertaken (Gordon & Weldon, 1973). The programming within these centers emphasized education and instruction, although there typically was an attempt to employ a "holistic" approach. Patients were encouraged to assume more responsibility for their own management (Noviks, King, & Spaulding, 1976). The families of the patients were frequently involved in the program as well. Interestingly enough, the psychological aspects of the disorder were apparently not emphasized to the extent one might expect. However, the support and learning that was brought about in those programs led patients to greater acceptance and control over their lives.

A day hospital program for treating obesity was described by Westlake *et al.* (1974). Their approach combined behavior therapy, group therapy, and family therapy in a 10-week part-time day center format. The program emphasized four major components: an information session, preparation of a meal, a behavior therapy session, and a family session. The information session focused on such topics as nutrition, proper diet, shopping habits, and the origin and effects of obesity. The authors stated that this session promoted group cohesion. The preparation and consumption of lunch was especially important in this program. Staff were used as models for eating habits and the patients who prepared the meal were evaluated on several dimensions relevant to their goals. After lunch, a 2-hour behavior therapy session was held. In this session, maladaptive habits were analyzed and new techniques were introduced to alter those behaviors. The family involvement centered on describing the program to relatives and enlisting their cooperation. The average weight loss in this program for 14 patients was found to be 14 pounds. A follow-up evaluation 6 months later revealed that these patients generally either maintained their weight loss or continued to lose weight. These findings were seen as being highly supportive of the program by the authors since "it is rare for weight loss to continue or even to be maintained after the completion of any supervised weight loss program" (p. 611). In light of the excellent results obtained in this program, it is surprising that the treatment of obesity in partial hospitalization settings has not become more commonplace.

Luber *et al.* (1976) described partial hospitalization services for patients presenting a primary diagnosis of epilepsy. These patients were integrated into a day program designed primarily for chronic psychiatric patients with severe deficits in interpersonal functioning and life management skills. The authors found the skill deficits of their epileptic patients to be comparable to those of their psychiatric patients. Programming for these individuals emphasized social skills training. Special consideration was given to daily living skills, conceptualization of ideas and feelings, and interpersonal relationships. The favorable treatment results with these patients led the authors to suggest that partial hospitalization "may afford epileptics a greater opportunity for normal integration into society" (p. 14).

Day programs specializing in the treatment of geriatric populations were first developed on a fairly large scale in England a number of years ago (Farndale, 1961). These centers serve both psychogeriatric populations and the physically infirm. Attendance at these facilities can vary from ½ day per week to a fully programmed 5-day week. The treatment emphasis is on occupational therapy, although activities aimed toward meaningful social interaction are typically stressed throughout the milieu. This type of specialized partial hospitalization program can frequently support individuals in the community who might otherwise be in psychiatric facilities or nursing homes (Goldstein, Sevriuk, & Grauer, 1968).

Farndale (1961) also described day programs in England that provide medical and industrial rehabilitation. Such rehabilitation centers have also appeared recently in the United States. These programs typically utilize physical therapy, vocational guidance, and occupational therapy as primary modalities. Most centers are oriented toward preparing their patients to assume full employment at some point in the future, although some actually provide longer term employment opportunities within their own facilities.

It may be seen, therefore, that partial hospitalization has been applied to a broad variety of treatment populations. Despite the great differences in settings and populations, partial hospitalization offers certain opportunities and advantages not readily available with other types of treatment models. A few of these unique advantages will be reviewed.

One characteristic common to all of the day programs that have been discussed is the emphasis on the patient's acceptance of personal responsibility. It is typical to find that partial hospitalization programming, regardless of the patient population, tends to encourage a sense of independence, self-reliance, and self-control. Also, the intensive cohesiveness and support in a day program can be quite helpful with a variety of problems. Day programs can provide experiences leading to the development of new coping skills. Experimentation with new ideas and skills is facilitated within a "safe," supportive environment. Yet partial hospitalization offers the opportunity to generalize what has been learned in the treatment setting to the patient's "real-life" setting on a daily basis. Being able to treat the patient in his own community frequently allows for a more natural, productive, and humane therapeutic experience.

These powerful characteristics of partial hospitalization account, in part, for the effectiveness of services noted in the diverse settings discussed. This potency, however, has not been applied to a number of populations that could clearly benefit from such an approach. Any type of medical, psychophysiological, or psychological difficulty that can be aided by changes in attitudes, increased emotional support, or new behavioral habits or skills could conceivably be treated effectively in a specialized partial hospitalization setting. Certain groups, in particular, may be especially benefited, such as cardiac patients, chronic pain sufferers, and "academic underachievers." Other patients who do not present typical psychiatric profiles but who possess similar deficits in skills of living could be integrated into existing day programs with much greater frequency, as Luber *et al.* (1976) demonstrated.

One factor that inhibits the expansion of partial hospitalization services into new or specialized areas is inadequate insurance coverage. Ironically, even though the utilization of partial hospitalization can bring about a significant dollar saving for the third-party payer (Goldberg, Chapter 9), there has been much reluctance to extend coverage to day programs. Typically, psychiatric day services cost approximately one-third to one-half that of inpatient treatment (Guilette, Crowley, Savitz, & Goldberg, 1978). Moreover, there is evidence that day treatment of certain specialized populations can bring about even greater dollar savings. In treating diabetics, for example, Spaulding and Spaulding (1976) found the cost of initiating insulin treatment to be *nine times greater* in the hospital than in their day program. With continued reports such as this, one hopes that insurance carriers will regard partial hospitalization with greater respect, understanding, and support.

In summary, partial hospitalization services have been extended to diverse psychiatric and nonpsychiatric populations. These specialized services have been developed within existing day programs as well as in separate facilities designed for particular purposes. It is anticipated that in future years, partial hospitalization will be creatively extended to a much broader scope of patients than is the current practice.

The Extended Patient Population: The Family

In discussing the relationships between the patient population and the treatment program, it is necessary to acknowledge that the identified patient is frequently a reflection of a troubled family system (Ackerman, 1958). Many partial hospitalization programs recognize this fact by including the families as part of an "extended" patient population. That is, the families are involved in some meaningful way in the treatment process. Many professionals believe that effective treatment is unlikely to occur if the pathological system surrounding the patient remains unchanged. Even those who do not employ such a "systems" approach generally appreciate the importance of involving family members in order to solicit their cooperation and support.

The manner and extent to which families are involved in the treatment process varies appreciably among day programs. Some centers initiate family contact by requesting or even requiring that all new patients be accompanied by a "significant other" to the intake interview. Families have also been brought into the treatment process through home visits (Peck, 1963). Frequently, family meetings are arranged shortly after the patient's entry into the program. In some instances, the family session is held without the patient, primarily for the purpose of obtaining information. At other times, the family and patient are seen together to assess the patterns of interaction. When the need for family intervention is indicated, many centers arrange for regular weekly family therapy sessions (Zwerling & Mendelsohn, 1965). When face-to-face contact with families is impossible, significant value can be obtained through periodic phone contacts (Odenheimer, 1965). Parkhurst and Hladky (1970) reported that the most pertinent family member was sometimes called in to spend a full day or more participating in the actual day program with the patient. MacKenzie and Pilling (1972) adopted a similar practice but went one step further by occasionally admitting the spouse as a full-time patient to the program.

Some writers have presented systematic approaches to family involvement in partial hospitalization settings. The multiple family group, for example, is one that has been applied with significant effectiveness. Davies, Ellenson, and Young (1966) described an open-ended multiple family group that sought to clarify the communications within the family, help families to define their own systems, and provide a setting where new and different ways of relating to one another could be explored. The authors found that the experience resulted in greatly improved communications within the family structure. Other benefits included positive changes in family roles, more effective handling of dependency relationships, and less scapegoating and special handling of the patient.

Luber and Wells (1977) described a structured, short-term multiple family group based upon a psychoeducational model. It consisted of seven 1½-hour sessions, each focusing on a specific technique or theme. The themes included such areas as patterns of expectation in the family, roles and role expectations, communications, empathic understanding, and multigenerational family patterns. Empirical data suggested that the group brought about positive changes in the families as well as in the identified patients. The authors concluded that the results, at least in part, were attributable to the structured, short-term nature of the group.

Moss and Moss (1973) presented a fairly comprehensive model for the involvement of families in a day program. They planned for both "direct" and "indirect" involvement of the patients' relatives. Direct participation included family sessions at the time of intake, family attendance at "house evaluation" of the patient's progress, and special meetings when treatment had "reached an impasse" (p. 170). In addition, an ongoing married couples group was held on a regular basis. In some instances, weekly family therapy sessions were arranged as well.

Indirect involvement centered around a weekly family discussion group for the members of the day program. In addition, a single-parents group was offered. Psychodrama was also frequently employed to help individuals gain a better understanding of family interactions. When patients had no family, they were encouraged to deal with the lost or fantasied family figures. The authors stated that integrating a "family orientation with group therapy of the day program provides a broad base for the treatment milieu" (p. 168). These authors have offered some helpful guidelines that should be consulted directly by those wishing to further the involvement of families in a partial hospitalization setting.

Therefore, the inclusion of families in the treatment process has taken many forms. Although systematic research has not been applied to many of these methods, family involvement certainly seems to represent a constructive program component. Compelling arguments exist for regarding families as an extension of the clinical population. It would appear that efforts in that direction should result in enhanced comprehensiveness and effectiveness of partial hospitalization programming.

Summary

This chapter has explored some of the complex interrelationships between partial hospitalization programming and the clinical population. Three major aspects of those interrelationships were emphasized: (1) the controversy surrounding the homogeneity or heterogeneity of the clinical population, (2) specialized patient populations, and (3) the involvement of families as an extension of the identified treatment group.

It is hoped that this chapter has stimulated a deeper appreciation of the relationships between treatment programming and the patient population. An understanding of the integral impact they have on one another can contribute to greater responsivity and relevancy in providing clinical services. Inadequate consideration of these influences, however, may lead to difficulties in fully comprehending the subtle and sometimes elusive "personality" of a partial hospitalization program.

References

Ackerman, N. *The psychodynamics of family life*. New York: Basic Books, 1958.

Astrachan, B. M., Flynn, H. R., Geller, J. D., & Harvey, H. H. Systems approach to day hospitalization. *Current Psychiatric Therapies*, 1971, 2, 175–182.

Beigel, A., & Feder, S. L. Patterns of utilization in partial hospitalization. *American Journal of Psychiatry*, 1970, 126, 1267–1274.

Davies, I., Ellenson, G., & Young, R. Therapy with a group of families in a psychiatric day center. *American Journal of Orthopsychiatry*, 1966, 36, 134–146.

Emerson, P., & Fagan, J. Integrating clients with mental retardation into comprehensive day treatment programs. In R. Luber, J. Maxey, & P. Lefkovitz (Eds.), *Proceedings of the annual conference on partial hospitalization*. Boston: Federation of Partial Hospitalization Study Groups, 1977.

Farndale, J. *The day hospital movement in Great Britain*. New York: Pergamon Press, 1961.

Glasscote, R., Kraft, A. M., Glassman, S., & Jepson, W. *Partial hospitalization for the mentally ill*. Washington D.C.: Joint Information Service, 1969.

Goldstein, S., Sevriuk, J., & Grauer, H. The establishment of a psychogeriatric day hospital. *Canadian Medical Association Journal*, 1968, 98, 955–959.

Gootnick, I. The psychiatric day center in the treatment of the chronic schizophrenic. *American Journal of Psychiatry*, 1971, 128, 485–488.

Gordon, P., & Weldon, K. The impact of diabetic day care centers on hospital utilization. *Nova Scotia Medical Bulletin*, 1973, *52*, 200–204.

Guilette, W., Crowley, B., Savitz, S., & Goldberg, F. D. Day hospitalization as a cost effective alternative to inpatient care: A pilot study. *Hospital and Community Psychiatry*, 1978, *29*, 525–527.

Harrington, J., & Mayer-Gross, W. A day hospital for neurotics in an industrial community. *Journal of Mental Science*, 1959, *105*, 224–234.

Hyland, J. The day hospital treatment of the borderline patient. In R. Luber, J. Maxey, & P. Lefkovitz (Eds.), *Proceedings of the annual conference on partial hospitalization*. Boston: Federation of Partial Hospitalization Study Groups, 1977.

Klein, J. The day treatment center for chronic patients: The politics of despair. *Massachusetts Journal of Mental Health*, 1974, *4*, 10–30.

Lamb, H. R. Chronic psychiatric patients in the day hospital. *Archives of General Psychiatry*, 1967, *17*, 615–621.

Laufer, M., Laffey, J., & Davidson, R. Residential treatment for children and its derivatives. In S. Arieti (Ed.), *American handbook of psychiatry* (Vol. 2). New York: Basic Books, 1974.

Luber, R. F. Adebimpe, V., & Trocki, D. *The role of partial hospitalization and social skills training in the treatment of epileptic patients*. Unpublished manuscript, 1976.

Luber, R. F., & Wells, R. Structured, short-term multiple family therapy: An educational approach. *International Journal of Group Psychotherapy*, 1977, *27*, 43–58.

MacKenzie, R., & Pilling, L. An intensive therapy day clinic for out of town patients with neurotic and psychosomatic problems. *International Journal of Group Psychotherapy*, 1972, *22*, 352–363.

Meltzoff, J., & Blumenthal, R. *The day treatment center*. Springfield, Millinois: Charles C. Thomas, 1966.

Moss, S., & Moss, M. Mental illness, partial hospitalization and the family. *Clinical Social Work Journal*, 1973, *1*, 168–176.

Noviks, L., King, B., & Spaulding, W. The diabetic day care unit. I. Development of an index to evaluate diabetes control. *CMA Journal*, 1976, *114*, 777–783.

Odenheimer, J. F. Day hospital as an alternative to the psychiatric ward. *Ardhives of General Psychiatry*, 1965, *13*, 46–53.

Ognyanov, V., & Cowen, L. A day hospital program for patients in crisis. *Hospital and Community Psychiatry*, 1974, *25*, 209–210.

Parkhurst, G., & Hladky, F. *The Tulsa day hospital project*. Final report, NIMH Grant MH-14700, 1970.

Peck, H. The role of the psychiatric day hospital in a community mental health program: A group process approach. *American Journal of Orthopsychiatry*, 1963, *33*, 482–493.

Spaulding, R., & Spaulding, W. The diabetic day care unit. II. Comparison of patients and costs of initiating insulin therapy in the unit and a hospital. *CMA Journal*, 1976, *114*, 780–783.

Washburn, S. L. Differentiation of treatment tracks for categories of patients. *PHSG Newsletter*, 1976 (Aug.), 3–4.

Weinstein, G. Pilot programs in day care. *Mental Hospitals*, 1960, *11*, 9–11.

Westlake, R., Levitz, L., & Stunkard, A. A day hospital program for treating obesity. *Hospital and Community Psychiatry*, 1974, *25*, 609–611.

Zwerling, I., & Mendelsohn, M. Initial family reactions to day hospitalization. *Family Process*, 1965, *4*, 50–63.

Zwerling, I., & Wilder, J. F. An evaluation of the applicability of the day hospital in treatment of acutely disturbed patients. *Israel Annals of Psychiatry and Related Disciplines*, 1964, *2*, 162–185.

9

Funding Partial Hospitalization Programs

F. Dee Goldberg and Joan Perrault

Introduction

Partial hospitalization for the treatment of psychiatric illness has been proven to be an effective treatment modality, yet its development in the United States has been far less than dramatic. Because of the demonstrations of clinical efficacy and cost-effectiveness of partial hospitalization, the lack of development of these programs is of concern. Although there are several reasons for the lack of development and utilization of partial hospitalization programs, this chapter will focus on those issues that are inextricably tied to financial considerations.

Financing Medical Care in the United States

The evolution of the present system of financing medical care has been confined to this century. As our nation has become more affluent and as developments in medical care and treatment have advanced, the American people have sought the opportunity to utilize the available medical services when necessary. During the administration of President Johnson, the availability of medical care was established as a right

F. Dee Goldberg • Deputy Commissioner, Program Support Service, Division of Mental Health, Columbus, Ohio 43215. **Joan Perrault** • George Mason University, and Department of Nursing, Mental Health Program Development, Fairfax, Virginia 22030.

of every citizen through the passage of Medicare and Medicaid legislation.

However, during the Depression, many people were unable to pay for the medical services they needed. Medical institutions striving to provide the highest quality of care were hampered by an insecure funding base. Patients were frequently financially incapacitated by unplanned illness, resulting in unemployment and high medical bills.

In the 1930s, the voluntary nonprofit health insurance industry was founded by people who wished to share the risk of large medical expenses. The program was developed in order to pay medical bills for those in need. The prepaid medical insurance plan had a board of directors, composed primarily of health care providers, which determined the limits of benefits paid on behalf of subscribers to providers. Contracts were made with hospitals for the cost of services plus a percentage of that predetermined cost, which was added for overhead expenses. The intention was to guarantee that services would not be denied to a beneficiary of the insurance plan based on ability to pay at a time of questionable financial security resulting from illness. This system enabled the institutions and physicians to provide services and receive reimbursement without confronting the consumer with the cost of the care.

As employee groups won the right to health insurance through collective bargaining, two things happened. First, since employers were paying the insurance premiums, the consumers of medical services became further removed from the payment of medical bills, thus losing any awareness of the actual costs of the services they had received. Second, the private for-profit insurance industry began to involve itself in the health insurance business. It offered employers a complete group insurance package, including life, health, and disability programs, thereby protecting their group insurance business. Many of these private plans provided payment either on a fixed per diem basis or on a coinsurance basis with the subscriber paying a percentage of the bill. If the per diem allowance was lower than the charges, the subscriber paid everything in excess of the allowance. If the subscriber paid a coinsurance, he paid a percentage of the total bill. In both instances, an awareness of the cost of medical care was maintained.

However, in Medicaid and Medicare legislation, institutional services were deemed a right, and it was determined that the consumer should pay a minimal portion of the bill, if any at all. Therefore, the majority of medical bills in the United States were paid by third parties, including Blue Cross/Blue Shield, Medicare, and Medicaid, on a cost-plus basis. This meant that employees covered by the private for-profit insurance plans, which had per diem limitations or significant coinsurance clauses, became dissatisfied when they compared their benefits with those covered under the cost reimbursement plans. The employers, therefore, began to modify employee benefit programs to provide the comprehensive benefits similar to those of the cost reimbursable plans.

Figure 1 illustrates the sources of payment of medical bills in the United States.

Financing Partial Hospitalization Programs

With this background, the financing of psychiatric care generally, and of partial hospitalization programs specifically, can now be addressed. As Figure 1 illustrates, insurance plays a less important role in the funding of mental health services than that of general health services. The development of insurance reimbursement for inpatient psychiatric units in general hospitals has had an adverse effect on the development of partial hospitalization programs. Treatment in psychiatric hospitals has frequently been excluded from insurance coverage or provided only very limited benefits; however, funding for treatment on an inpatient psychiatric unit of a general hospital has been the same as that for the medical or surgical units of the hospital, creating a bias

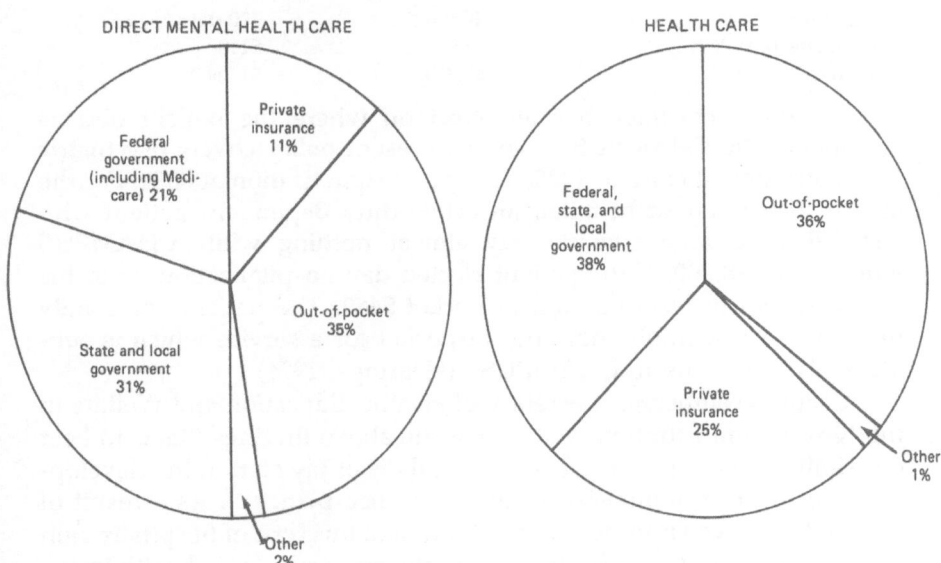

Figure 1. Sources of funds. Left, from *The Cost of Mental Illness*, by Dan and Dianne Levine, DHEW Pub. No. ADM 76-265, 1971; right, from *The Size and Shape of the Medical Care Dollar*, DHEW Pub. No. SSA-11910, 1973.)

toward treatment in these units. These factors favor treatment in an inpatient psychiatric unit located in a general hospital in the community, where partial hospitalization programs should be located. Since insurance coverage of services at a general hospital frequently provides for 100% of the cost of inpatient treatment and requires a coinsurance factor for outpatient services, sources of funding for partial hospitalization are limited.

This funding delemma was presented to Congress by S. Alan Savitz, M.D. (previously the medical director of the Silver Spring Day Treatment Center, Silver Spring, Maryland), in his testimony on CHAMPUS benefits. (CHAMPUS is the Civilian Health and Medical Program of the Uniformed Services and is a system for financing hospital and outpatient services provided to active duty military personnel, those retired from military service, and their dependents.) A report of the testimony is quoted in full for clarification of this issue.

"Translating the above into dollars and cents, a typical patient in the Washington area would be faced with the following costs:

	Day Hospitalization	Inpatient Hospitalization
Average length of stay (days)	80 days	120 days
Treatment (days billed for)	40 days	120 days
Facility fees (per day)	$55	$149
Total facility costs	$2,200	$17,880

Insurance coverage has an effect on where the patient obtains treatment. The CHAMPUS programs present policy covers psychiatric hospitalization at close to 100% and day hospitalization at 80%. For the stay in the inpatient hospital, an active duty dependent patient with CHAMPUS coverage would pay almost nothing while CHAMPUS would pay $18,000. If the patient elected day hospitalization he or his family would have to pay out-of-pocket $480. The patient and family must pay considerably more out-of-pocket for a service which is substantially less costly to CHAMPUS" (Hearings, 1974).

Casper Weinberger, Secretary of Health, Education and Welfare in the Nixon administration, stated that the above findings "tend to bear out similar conclusions reached by members of my staff in the development of our comprehensive health insurance plan. . . . As a result of this we have given appropriate recognition to this type of hospitalization by providing mental health benefits in the comprehensive health insurance plan that would cover 60 days of partial hospitalization during each year, or double the inpatient coverage of 30 days" (Hearings, 1974, p. 421).

In the private sector, the existing financial incentives are directed toward inpatient care. Two partial hospitalization programs in the metropolitan Washington area worked intensively with Aetna Insurance Company to reverse this trend. In a presentation to the National Association of Private Psychiatric Hospitals, Malcolm MacIntyre, Aetna's Government Relations Director, described a pilot program that provided benefits for partial hospitalization equal to those for inpatient care.

The Aetna pilot project was developed as an answer to extremely aggressive complaints from the staffs at the Silver Spring Day Treatment Center and the Potomac Foundation for Mental Health, two private psychiatric day hospitals in the metropolitan Washington area. Aetna's cutback in mental health benefits under the Federal Employee Health Insurance Program initiated this strong and persistent reaction. In 1974, the Aetna Insurance Company was experiencing extreme financial pressure because of excessive utilization of psychiatric insurance benefits in the metropolitan Washington area. In fact, the percent of total benefits being utilized for nervous and mental disorders rose to 25% of total benefits. A study of the claims paid indicated that inpatient benefits utilized two-thirds of mental health expenditures. Aetna felt that utilization controls on inpatient services were sufficient but that outpatient services were not appropriately reviewed or controlled. They therefore negotiated with the Civil Service Commission to change the outpatient psychiatric benefit. The new contract limited the number of outpatient visits to 20 in a private practitioner's office and to 40 in an organized mental health center. Staff from the two psychiatric day hospitals carefully explained, although with quite strong dialogue at times, that the change in the Aetna contract, classifying partial hospitalization as an outpatient service and limiting the number of outpatient visits, was not only irresponsible and based on a lack of understanding but also expensive. In fact, it would have cost Aetna approximately $250,000 more based on the limited experience of patients with Aetna coverage treated at the Silver Spring Day Treatment Center over a period of 3 years. It was further stated that if those officials who were making decisions and representing Aetna in discussions with providers were not interested in discussing this costly decision, the president of Aetna might be interested in finding out how Aetna's cost containment program resulted in unnecessary expenditures approaching a quarter of a million dollars. The Aetna officials agreed to meet and discuss the issues with representatives of the Silver Spring Day Treatment Center and the Potomac Foundation for Mental Health.

During future negotiations, representatives of the partial hospitalization programs provided Aetna officials with diagnostic and treatment statistics, which indicated that the programs did, in fact, act as alternatives to inpatient care and not as intensive outpatient therapy. Each Center was accredited by the Joint Commission on Accreditation of Hospitals and participated in a model external utilization review program. Thus both facilities demonstrated intensive treatment programs, which included utilization review and national certification.

The pilot project was begun under special criteria detailed in the MacIntyre speech cited above. In a preliminary evaluation of the pilot project results, Dr. William Guillette, medical director, Group Claims Division of Aetna, summarized the findings. The pilot included 17 patients at a total treatment cost of $64,000. The estimated cost of treating these patients in a hospital was $178,000 for a net saving of $114,000, or a saving of approximately two-thirds (Guillette, personal communication).

This estimated saving does not illustrate the efficiency of day hospitalization demonstrated by Herz, Endicott, Spitzer, & Mesnikoff (1971). In a controlled study, the authors compared the length of stay and readmission rates of patients treated in a partial hospitalization program and in an inpatient setting. Findings showed the average length of stay for inpatients was 119 days while day patients had an average stay of 49 days, considerably less than half that of inpatients. Readmissions for day patients was 22%, compared with 42% for inpatients, nearly twice that of day patients. Thus, had these phenomena been included in the interpretation of the Aetna study, the net saving would have approached 80%.

There are several factors that contribute to the clinical efficacy and therefore the cost-effectiveness of psychiatric partial hospitalization (Goldberg & Goldwater, 1977):

1. It discourages the excessive dependence and regression that often occurs with inpatient hospitalization, especially intermediate and long-term hospitalization.
2. It avoids the isolation and dehumanization of inpatient hospital facilities.
3. It encourages higher expectation levels for each patient because patients must maintain all of those independent activities of which they are capable, despite their mental illness. As has been frequently demonstrated, patients and families, most times, do respond to expectations and structure regardless of the severity of the emotional psychopathology.
4. The patient remains within the family, which forces the patient and family to work through family problems.
5. Flexibility of the schedule allows the program to meet the patient's needs, decreasing the frequency of attendance as the pa-

tient reintegrates into the community. The schedule approximates the work week so that return to employment is facilitated even as the patient continues in treatment.
6. There is much less social stigma than with inpatient hospitalization.
7. Fewer staff must interact with the patient and each other making information processing and treatment planning more effective.

Barriers to Financing Partial Hospitalization

The way in which insurance funding discourages the development of partial hospitalization programs in general hospitals has already been discussed. In publicly funded programs, however, the phenomenon of double funding acts as a barrier. In order to develop an alternative program to an existing one, it is necessary to provide duplicate funding during the period of transition, unless one program ends precisely at the time the alternative program begins. For example, a partial hospitalization program may be established in the community for a fraction of the cost of providing services in an inpatient setting, but if a number of beds are not closed when the partial hospitalization program is started, the cost of both services continues to be incurred, and no savings will accrue. In other words, it is essential that partial hospitalization programs replace some inpatient beds. The findings of the Herz *et al*. (1971) study support the feasibility of this recommendation. As long as there is more comprehensive insurance coverage for inpatient care than for partial hospitalization, the incentive to make the necessary changes is virtually nonexistent.

A second major barrier to the financing of psychiatric care is composed of both the difficulty within the mental health field for mental health practitioners to set standards of care and the rigidity required in the philosophy of insurance findings. Insurance is based on the principle of risk; that is, of the entire population of individuals enrolled, a small number, predicted by experience, will require payment of benefits in a given benefit period. Since the criteria for the diagnosis and treatment of various mental illnesses vary widely from institution to institution and provider to provider, it is difficult to get sufficient data to predict risk and cost of care. As a result, insurance coverage for the treatment of mental illness may be compromised, especially if that treatment occurs outside of an inpatient setting where third-party payers feel some utilization controls exist.

It has been illustrated that partial hospitalization is less costly than inpatient services. For providers reimbursed on a cost-plus basis, there is no incentive to decrease costs. If, for example, the overhead allowance (or "plus" factor) is 10% and inpatient care costs $150 per day, the total reimbursed would be $165 per day with $15 being the overhead allowance. If a partial hospitalization program costs $40 per day, the total reimbursed would be $165 per day with $15 being the overhead allowance. If a partial hospitalization program costs $40 per day, the total service reimbursement, the total dollars received by the institution are based on charges for each service, so if the charges are lower (as for partial hospitalization) the total receipts are lower. Again, the bias from the institutional viewpoint is against partial hospitalization.

A final factor involved in the resistance to the development of more partial hospitalization programs deserves mention because of its political, economic, and sociological ramifications. If, as has been suggested, the creation of partial hospitalization programs is paralleled by a decrease in inpatient beds, there will be a need for fewer employees, at a ratio of 1:4.2. This ratio is derived from the fact that most partial hospitalization programs require a staffing pattern for a 40-hour week, composed of 5 8-hour days and an inpatient setting requires staffing coverage for 168 hours a week (40/168 = 1:4.2) The need for more staff for an inpatient setting for 7 days a week, 24 hours a day is evident, although during much of this time no active psychiatric treatment occurs.

Summary

Several of the partial hospitalization programs currently in existence were started with grants under the Community Mental Health Center legislation with the expectation that reimbursement from third-party payers would eventually support these programs. So far this has not occurred. For this reason, and also because intensive care in partial hospitalization programs is not funded in the same way it is on an inpatient unit, many of the partial hospitalization programs become aftercare, or maintenance programs without the necessary trained staff to run a short-term acute care service.

As our country moves in the direction of a national health policy, the time is ripe for some changes in the means of financing psychiatric care. It is important that insurance contract administrators, union leaders, personnel directors, insurance company executives, federal, state, and local legislators, and health planners become advocates of partial hospitalization programs. To reach these individuals it is helpful to develop a one-page fact sheet detailing the commonsense benefits, both clinical and financial, of partial hospitalization programs. Often the strongest advocates of partial hospitalization programs are people who have had personal experience with an individual who has had the benefit of treatment in a partial hospitalization program. Union leaders and personnel directors who have seen employees return to work a short time after experiencing a serious emotional illness, and parents whose children have had a serious emotional illness and returned rapidly to normal levels of functioning, can provide decision-makers with the information they need about the benefits of partial hospitalization. Mental health practitioners must provide these potential advocates with a clear understanding of how partial hospitalization is so effective and with fact sheets and other educational tools to maximize their effectiveness in educating policy-makers.

If a national health security program is enacted that provides a per capita expenditure allowance, local control of planning and allocation, and a systems approach to the development and support of health services, programs such as partial hospitalization services, which are both clinically effective and cost-effective, will naturally evolve. If, however, a national health insurance program is built on the existing funding mechanisms which support traditional services, advocates of partial hospitalization programs will have an even greater task establishing recognition for reimbursement for services. The commitment of a national health insurance program to traditional institutional services that currently receive favorable funding would continue and could totally consume all available health services funding.

References

Goldberg, F. D., & Goldwater, M. Obtaining state legislation for insurance coverage of day hospitalization. *Hospital and Community Psychiatry*, 1977, *28*, 448–450.

Hearings on CHAMPUS and Military Health Care before a subcommittee of the Committee on Armed Services, House of Representatives, 93rd Congress, 2nd session, October 11, 1974.

Herz, M. I., Endicott, J., Spitzer, R. L., & Mesnikoff, A. Day versus inpatient hospitalization: A controlled study. *American Journal of Psychiatry*, 1971, *127*, 107–118.

Future Directions of Partial Hospitalization

Raymond F. Luber

Introduction

The previous chapters in this book have considered a wide spectrum of issues related to partial hospitalization. Programmatic treatment approaches, special patient populations, research issues, and problem areas have been considered in detail. Individually and in combination these chapters highlight the significant conceptual, clinical, and research factors involved in partial hospitalization as it is generally utilized at the present time. In addition, the pragmatic and difficult question of funding has been explored.

Many questions remain unanswered, however. What kinds of changes do partial hospitalization services effect in patients? What particular techniques, groups, or treatment orientations most effectively produce these changes? For whom is the treatment modality most beneficial? How can the aims and functions of partial hospitalization be most effectively defined and implemented? What sources of funding can be generated and how can this be done most effectively?

In some respects these issues do not differ dramatically from those faced currently by many forms of psychiatric treatment (e.g., group psychotherapy, Parloff & Dies, 1977; parental training, Forehand & Atkeson, 1977) for the empirical-scientific approach has a relatively short

Raymond F. Luber • Department of Psychiatry, Western Psychiatric Institute and Clinic, University of Pittsburgh School of Medicine, Pittsburgh, Pennsylvania 15260.

history in psychiatry. Doubtless, some of the important questions regarding partial hospitalization appear more immediate than others, depending on the perspective from which they are viewed. Administrators, already confronted by dilemmas created by funding and cost-effectiveness issues, may, for example, view the more global issues of
treatment orientation and appropriate patient selection as of less than
immediate importance. On the other hand, clinicians facing daily treatment decisions and responsibilities may be primarily concerned with
issues related to the question: "What treatment works?" And finally,
researchers interested in exploring more definitive areas may be anxious
to focus on issues designed to clarify the question of the maximally
appropriate patient for partial hospitalization treatment.

Initially, it may appear tempting to focus on one of these issues or
perspectives as singularly important. Such a concentration would be a
major error, however, for all three perspectives are crucial to the future
of partial hospitalization. This chapter will summarize the current status
of partial hospitalization and delineate the major issues that will affect
the future directions of the treatment modality.

Research

The research area will be considered first both because it is an issue
of major importance in its own right and because it is integrally related
to many other issues bound to be of major significance in the future
development of this treatment modality. It can be accurately stated that
at the current moment partial hospitalization has marshaled a significant
amount of conceptual evidence to establish its face-validity as an effective treatment modality. Contributors to this volume, however, have
repeatedly pointed out the general absence of empirical evidence to
support much of this conceptual framework. This, of course, is not to
say that partial hospitalization is a completely undocumented treatment
approach. As Hersen (Chapter 6) has described, research has been conducted in a number of significant areas: patterns of utilization, program
evaluation, comparison with inpatient hospitalization, comparison of
different treatment approaches, and the investigation of individual patients in the partial hospitalization setting utilizing the single-case research design. In addition, the literature describes increasing instances
in which validated treatment techniques are being utilized within partial
hospitalization programs (e.g., Hersen & Bellack, 1976; Hersen & Luber,
1977; Liberman & Smith, 1972; Lombardo & Turner, in press; Luber &
Hersen, 1976; Williams, Turner, Watts, Bellack, & Hersen, 1977).

Nevertheless, at this point, partial hospitalization seems to have reached a major crossroads in its development. Stated in an admittedly oversimplified manner, the future of partial hospitalization will be either in a direction characterized by the accumulation of empirical data or in the direction of unsubstantiated treatment and continued reliance on the apparent face-validity of the modality. Obviously, the choices are not dichotomous or so clearly defined. Treatment in partial hospitalization programs cannot be discontinued while awaiting the resolution of all unanswered questions for, regardless of the situation, the number of programs and the number of patients treated in these programs is steadily increasing (see Chapter 1). Nevertheless, many other pragmatic issues faced by partial hospitalization programs are directly related to a continued and increased emphasis on research investigation.

For example, the proliferation of systems of utilization review will soon expand to outpatient and, inevitably, to partial hospitalization programs (Wollowitz, 1977). Such systems seek to ensure the provision of quality treatment at the lowest possible cost. They include stringent and detailed documentation of intake and treatment procedures as well as of treatment results and discharge planning. Thus, to satisfy pending utilization review requirements and third-party payers and, consequently, to maintain a patient population, partial hospitalization programs will be required to substantiate their treatment.

Secondly, the area of financial support is integrally related to continued emphasis on research. As Goldberg and Perrault have indicated (Chapter 9), partial hospitalization is currently caught in a crossfire between institutional and third-party payer biases favoring financial support of inpatient treatment. However, the continually increasing costs of inpatient psychiatric care coupled with the initially encouraging findings regarding the utility of partial hospitalization in comparison to inpatient treatment may alter this picture somewhat. But regardless of these developments, it appears certain that partial hospitalization programs will be called upon to demonstrate their efficacy if funding sources are to be maintained and expanded, whether these be public or private sources of revenue.

On two counts, then, the future of partial hospitalization must be in the direction of empirical investigation. First, efforts must be directed toward determining the *efficacy* of this particular treatment modality. These efforts will undoubtedly be in some of the areas suggested by Hersen (Chapter 6). However, one aspect seems of particular interest and importance, namely, determining the utility of validated treatments in the *context* of partial hospitalization. Most partial hospitalization programs are a combination of various treatment ingredients. The interactive effect of these treatments has yet to be investigated. For example, what effect occurs when systematic desensitization or social skills training are presented within the framework of a total partial hospitalization program? Do other program components accelerate, retard, or have no effect on the specific treatment applied? Can other program components be used to test or enhance the generalization of particular treatments (e.g., social skills training)? Such questions are unanswered at this point.

Another aspect of interest in this area involves the efficacy of various treatment orientations or approaches. In this volume two specific treatment frameworks have been described, namely, the behavioral and the milieu approaches. In each instance, programmatic implications and treatment results have been presented. In general, these procedures have been supported by positive treatment results. At the same time, another investigation has offered some *minimum* indications that the behavioral approach offers positive advantages over other types of treatment (Austin, Liberman, King, & DeRisi, 1976). Nevertheless, no evidence exists to substantiate the utility of *either* orientation in more specific terms (i.e., is either approach more effective with specific diagnostic categories, behavioral characteristics?, etc.). The investigation of this area should be of major interest for partial hospitalization in the future.

Second, research efforts must be directed toward the resolution of *recurrent questions* and problems that have faced partial hospitalization almost since its inception. These areas will be considered briefly below.

Program Function

What is the function of a partial hospitalization program? Should it serve as an alternative to full-time hospitalization, as a transition to full community living, as a rehabilitation setting, or as a service designed for distinct populations? Various studies and conceptual formulations previously considered have indicated that partial hospitalization may provide a useful service in the performance of each of these functions. This, in turn, has increased the enthusiasm of partial hospitalization advocates and encouraged expansion into all of these areas. To some extent, a fragmentation and dilution of services has resulted from this expansion.

In light of funding difficulties and the general underutilization (see below) encountered by partial hospitalization programs, this diversification is understandable, although perhaps unfortunate. As has been effectively pointed out previously, it is highly unlikely that a *single* program can encompass all of these functions without seriously impairing its treatment effectiveness. Since the execution of each of these functions requires somewhat different treatment emphases, it is difficult to envision their incorporation into a single program.

For example, the distinctive functions of supplying an *alternative* to hospitalization and providing a *rehabilitative* program for chronic patients may be incompatible inasmuch as treatment approaches would differ greatly. A program organized to provide an alternate to hospitalization would, undoubtedly, focus on acute patient problems. Program emphasis would thus include an intensive assessment of factors such as the patient's level of functioning, interactions with others, and potential for engaging in destructive activities (aimed at self or others). In addition, a heavy emphasis on pharmacotherapy would probably be indicated. Finally, treatment in this type of program would probably concentrate on helping the patient regain the ability to function actively in the community and family environment, and would be carried out by a clinically sophisticated staff of professionals.

A rehabilitative program for chronic patients would, on the other hand, have a distinctly different focus. Here, resocialization and reeducation would be the main emphasis. More limited goals might be identified in this type of program (e.g., teaching skills related to the activities of daily living versus preparing for return to full-time employment). Though the administration of psychotropic drugs would be a fundamental aspect of such a program, the major emphasis might be on *maintenance* of prescribed drugs in light of the chronic patient's tendency to discontinue medications prematurely. Finally, a more diversified staff with specialized training in certain areas (e.g., recreation) might be utilized in a rehabilitative program.

From this brief example, the difficulty of combining programs with two distinct functions can be seen. In this specific instance, the difficulty is further illustrated by the decreasing tendency to combine inpatient and partial hospitalization programs despite the obvious economic advantages inherent in this arrangement.

One significant question that must be answered in the future, then, involves the most effective utilization of partial programs to fulfill the functions that have been conceptualized for the treatment modality. Two solutions seem possible: A single program may develop various treatment components to satisfy these multiple functions; or, several programs acting in concert may agree to jointly provide the needed services by each assuming distinct functions.

Both alternatives pose significant difficulties. In the first instance, the cost of maintaining comprehensive partial programs could be prohibitive; also, in areas of limited population, such concurrent programming may be totally impractical. On the other hand, securing the cooperation of several programs and agreeing on the division of functions might be equally difficult; in addition, if these programs were dispersed over a large geographical area, transportation problems, similar to those already faced by many rural programs, would have to be resolved.

Perhaps no single solution to this problem exists. Lahey (Chapter 4) has called for flexibility in the treatment of children and adolescents, with the sharing of treatment responsibility among family, community, and institution. The most efficient utilization of partial hospitalization programs may no longer be to maintain isolated, self-contained treatment programs and goals. A voluntary, cooperative effort in which the needs of patients, the resources available, and the differential functions of partial hospitalization are identified and considered in the provision of services may be the only feasible avenue to pursue. As evidence accumulates regarding the most appropriate functions of partial hospitalization, those individuals organizing and administering such programs may be required to revise their attitudes and operating procedures in order to establish a *primary* task of their program.

Patient Population

Some 15 years ago the selection of appropriate and suitable patients for partial hospitalization was identified as the key problem to be faced in the future (Zwerling & Wilder, 1962). That issue is as yet unresolved. With the increased emphasis on deinstitutionalization and non-institutionalization (i.e., the least restrictive form of treatment), the question "Who are appropriate patients for partial hospitalization?" is of even greater importance. Partial hospitalization programs are now being called upon to treat many who have previously been considered appropriate for long-term institutional care but who are currently being released into the community (Krieger, 1977). In addition, those who might previously have been regarded appropriate for short-term institutional care are now being recommended for partial treatment instead.

Furthermore, as Lefkovitz (Chapter 8) has pointed out, partial hospitalization has been utilized in the treatment of a broad spectrum of psychiatric and nonpsychiatric patients including chronic schizophrenics, borderline personalities, geriatrics, diabetics, epileptics, and those with problems of obesity. Finally, partial hospitalization programs for children and adolescents as well as the mentally retarded have been initiated and are described in this volume.

The conclusion emerging from the above is that almost any general patient classification has been treated within the context of partial hospitalization. However, to date very little hard data exist identifying those variables that will predict success or failure of treatment. We are not able to say *which* chronic schizophrenics, adolescents, or mentally retarded individuals, for example, will benefit most from partial hospialization treatment. Nor are we able to determine with any degree of certainty whether partial hospitalization is *more effective* with chronic schizophrenics than with depressives, than with borderline personalities, etc.

As has been indicated previously, some general conceptual parameters regarding the most appropriate partial hospitalization patients do exist. With adolescents such indicators as childhood schizophrenia and the strength of the child (i.e., the amount of environmental support and encouragement received by the child) have been proposed as possible predictors of suitability for and success in partial hospitalization programs.

Another general indicator has been suggested by Beigel and Feder (1970), namely, chronicity. Essentially, this investigation indicated that diagnosis, suicide potential, and family involvement were *not* predictive of full utilization of partial hospitalization. On the contrary, the only predictive factor was chronicity. That is, those patients admitted with acute illness or an acute exacerbation of chronic illness had a significantly greater chance of complete utilization of the partial hospitalization program than those entering to modify a long-standing, undesirable aspect of their behavior or life-style. It must be remembered, however, that this study applied to both acute and chronic patients treated in the *same* program. This does not mean that chronic patients cannot be effectively treated in partial hospitalization, but as the authors themselves conclude, separate programs may be necessary for the two categories of patients.

A further significant consideration in regard to the issue of patient population is the indication that, for some patients, partial hospitalization may be a regressive treatment form (Herz, 1977). This is especially true in instances in which partial treatment is utilized as a substitute for work, school, or vocational training. The crucial element in this situation involves the ability to accurately assess patient potential and determine at what point continued hospital treatment should be discontinued in favor of a return to more normal community and family involvement. This determination, in turn, has implications for the selection of suitable partial hospitalization patients. It represents a second perspective somewhat different from that of "Who will *benefit* most from the treatment?" That perspective is, of course, "What patients may be *adversely* affected by inclusion in partial hospitalization treatment?"

A final complicating factor involved in patient selection is the finding of several studies (Fink, Longabaugh, & Stout, 1977; Fink, Heckerman, & McNeil, 1977; Grob & Washburn, 1977; Washburn, Vannicelli, & Scheff, 1976) that clinical characteristics traditionally utilized for determining treatment modality or setting (e.g., diagnosis, presenting symptoms) are of limited utility in determining appropriateness/inappropriateness for partial hospitalization. These studies will be considered in greater detail below. It is sufficient to say here that in each case the *clinical* presentations of inpatients and day patients did not differ significantly. What then should determine appropriate treatment? One suggestion (Fink, Heckerman, & McNeil, 1977) is that the only clinically valid reasons for inpatient admission may be to control destructive behavior and/or to implement an effective treatment plan through 24-hour supervision. If future investigations should substantiate this hypothesis it will be increasingly important to delineate *contraindications* for partial treatment as well as indications for inclusion in this form of treatment.

To date, initial efforts have been made to explore the question of suitability of patients for partial hospitalization. Definitive criteria, however, have not yet been established which can be effectively utilized in screening and selecting suitable admissions to programs. The development of such criteria would have self-evident utility in assisting in the provision of maximally effective treatment and should be a major emphasis in the future development of partial hospitalization.

Underutilization

Despite the increasing evidence supporting the clinical utility of partial hospitalization as an alternate to 24-hour hospitalization, as a transitional treatment facility, and as a fiscally effective treatment modality, these programs have experienced some difficulty establishing themselves in the continuum of modern psychiatric care. In short, partial hospitalization programs often suffer from a problem of input (i.e., they receive an inadequate number of referrals) and consequently are significantly underutilized.

Previous investigations have indicated that several factors other than clinical issues are involved in final determinations regarding hospitalization and/or treatment. For example, Hollingshead and Redlich (1958) demonstrated that various nonprofessional, nonclinical, and nonscientific factors were significantly important in determining the type of hospital a patient entered as well as the form of treatment he/she received. Since then, Mischler and Waxler (1963), Mendel and Rapport (1967), Maxmen and Tucker (1973), and Lubin, Hornstra, Lewis, and Bechtel (1973) have expanded on these findings and demonstrated that socioeconomic, demographic, interpersonal, attitudinal, and organizational factors are important nonclinical determinants involved in treatment decisions.

Feeling that similar considerations might be operating in the under-utilization of partial hospitalization, several investigators have recently studied the problem. These studies, their results and implications will be briefly reviewed.

Washburn *et al.* (1976) observed 392 women admitted to a psychiatric hospital in order to determine the relative value of full-time hospitalization versus day hospital treatment. Four categories were utilized to classify treatment recommendations: (1) too well for day treatment, (2) too sick for day treatment, (3) suitable for day treatment, and (4) feasible for day treatment. The study revealed that although 59% of the sample was suitable or feasible for treatment in partial hospitalization, only 21% was actually able to exercise that option. The investigators found several factors acting as deterrents to treatment in partial hospitalization, including inadequate insurance coverage or private funds (20%), failure of the patient or relative to give permission for the patient's participation (16%), and failure on the part of clinicians to make a final decision to pursue partial hospitalization treatment (15%).

In both of the latter instances, staff *attitudes* were found to be significant in treatment decisions. In particular, first-year psychiatric residents were found to strongly emphasize possible problems that might be encountered in partial treatment, while more experienced physicians were found to be unable to accept partial hospitalization as an adequate modality for "intensive" treatment. The authors concluded that the underutilization of partial hospitalization "has been based as much on difficulties imposed by personal biases or unfamiliarity with such facilities as by psychopathology and financial realities of patients" (p. 182).

In a second study, Grob and Washburn (1977) compared patient characteristics in inpatient and day hospital settings and examined the factors involved in the selection of treatment facilities. A sample of 90 patients was used consisting of three groups: (1) Day (admitted directly to the day hospital from the community), (2) In-Day (admitted to the day hospital in transition from the inpatient unit), and (3) In (admitted directly to the inpatient unit). The In and Day groups were matched for age, sex, and diagnosis.

Data were examined in terms of five of the six categories enumerated by Maxmen and Tucker (1973), with organizational factors being excluded. Using these categories, the only *socioeconomic* or *demographic* factor differentiating the groups involved financial resources: Patients admitted to inpatient services had insurance benefits for hospitalization while partial hospitalization patients did not have similar benefits for partial hospitalization and were financed by alternate funds. One *interpersonal* factor distinguishing the groups was the fact that significantly more inpatients experienced dramatic environmental changes (death, accident, loss, or family stress) just prior to hospitalization. *Clinically*, partial patients were found to have disturbances as severe as those found on inpatient services. And finally, *attitudinal* characteristics of clinicians revealed by the study as influencing the hospitalization decision included (1) severity of symptoms (more important for In sample admissions), (2) need for a structured program (more important for In-Day and Day groups), (3) failure to consider referral prior to the actual time of referral, (4) expectation of a longer length of stay in partial hospitalization, and (5) familiarity and/or experience with the treatment facility to which the patient was being referred.

Of particular interest in this study, as it relates to the underutilization of partial hospitalization, is the fact that, taken as a whole, attitudinal factors do influence admission decisions. It appears that some factors (e.g., expectation of a longer treatment stay in day hospital, failure to preplan referral, and familiarity with treatment facility) might work against partial hospitalization and in favor of inpatient treatment. They also imply that some educational efforts regarding the nature, goals, and functions of partial hospitalization may be useful in increasing the utilization of the treatment modality. This seems especially true in light of the fact that referrals to inpatient treatment were based on the severity of the patient's symptoms when, in reality, the symptom severity of inpatients and day patients were not significantly different.

A third study by Fink, Longabaugh, and Stout (1977) adds further data for consideration. In this instance, two samples of 40 patients each drawn from inpatient and day hospital admissions were compared. The two groups showed no statistically significant differences in age, sex, diagnosis, severity of illness at admission or discharge, or history of previous inpatient admissions. Outcome measures were completed on the basis of a personal interview with each patient and a "significant other" and included the Psychiatric Evaluation Form, a Subjective Distress and Role Functioning Scale, the Butler Hospital Mental Status Form, and a scale adopted from the International Study of Schizophrenia. Severity of impairment in the problem areas identified in the original admission problem list was also rated. A 1-year follow-up showed no statistically significant differences between the two groups on any of these measures. In addition, although treatment costs during the 1-year follow-up period were not significantly different for the two groups, the median cost of the *initial* hospitalization for partial hospitalization patients ($1,351) was significantly lower than that of the inpatient sample ($3,245), thus resulting in a significant difference in Total Treatment Costs (Day Hospital: $2,331; Inpatient: $4,156).

Given the clinical effectiveness and cost-saving characteristics of partial hospitalization, why is it underutilized? One reason uncovered in this study was the "reluctance of patients, clinicians, families and third party payers to support the treatment of given patients in the partial hospitalization setting" (p. 92). That is, although many in the study sample were clinically suitable for treatment in partial hospitalization, this was not considered an acceptable alternative for inpatient hospitalization by patients and their families.

A second reason for failure to consider partial hospitalization revealed in this study related to attitudinal characteristics of admitting physicians. Specifically, although physicians saw patients as clinically suitable for partial hospitalization, they argued that inpatient treatment was necessary in order to separate the patient from the family and thus "learn more about him" and provide "intensive" treatment. However, considering the results obtained at the 1-year follow-up, the investigators accurately point out that it appears this separation provided little information useful in improving clinical outcome.

Again, this investigation provides initial data indicating that nonclinical considerations and biases are quite important in the making of treatment decisions. Indeed, these biases seem to be as important as the severity of psychopathology and financial resources in the determination of treatment setting.

Finally, a study reported by Fink, Heckerman, and McNeil (1977) bears on the nonclinical factors involved in the underutilization of partial hospitalization. Based on their impression that subjective factors involved in treatment setting decisions have a detrimental effect on referrals to partial hospitalization, these investigators examined the medical records of all patients referred to either inpatient or partial hospitalization treatment for a 1-year period at a private, nonprofit, university-affiliated hospital. This examination revealed that 46% of all referrals to partial hospitalization during this period were accounted for by 2 of 10 physicians responsible for the determination of the most appropriate treatment site for all patients evaluated at the hospital. Of these 46%, 1 physician (the program director of partial hospitalization) accounted for 33% while the other physician (the original organizer of the partial hospitalization program) accounted for 13% of the referrals. Further, it was determined that patients were more likely to be referred to the partial hospitalization program by (1) self, family, or friend, (2) public psychiatric hospitals, and (3) outpatient mental health clinics.

In an effort to document the absence of clear-cut clinical differences between patients treated in the two settings (i.e., inpatient and partial hospitalization), Fink et al. (1977a) investigated 52 clinical and symptom items for all admissions during this 1-year period. In regard to *diagnosis*, results indicated that this classification alone did not appear to be of great utility in helping select the most appropriate treatment setting for patients. In addition, with the exception of a greater occurrence of phobias and obsessive/compulsive symptoms in the partial hospitalization group, other *sign and symptom* profiles did not differentiate the two groups in any way useful for determining treatment setting. Finally, aside from the finding that partial patients tended to be younger, unmarried, and living in the parental home, no *sociodemographic* or *psychiatric history* variables emerged as clinically relevant in assisting in the assignment of patients to particular treatment settings. In sum, the investigators concluded that clinical differences that exist between inpatient and partial hospitalization groups would not systematically aid in determining the appropriate treatment setting for patients. On the contrary, it was their opinion that "a major variable in the treatment setting selection process is the personal characteristics (i.e., clinical biases) of the individual who makes that clinical decision" (p. 9) and that this bias "is at least as important as is a patient's specific clinical characteristics" (p. 10).

In summary, then, several investigations have generated preliminary findings elucidating the underutilization of partial hospitalization. Included in this category are several nonclinical factors such as insufficient insurance coverage (Washburn et al., 1976), attachment to preexisting mental health services (Beigel & Feder, 1970), lack of awareness on the part of potential referral sources of the existence and/or purposes of partial hospitalization programs (Finzen, 1974; Fotrell, 1973), transportation difficulties (Bierer, 1959), and other clinician-specific subjective biases suggested in the studies cited above. These biases include the opinion that "intensive treatment" could best be provided in an inpatient setting, the feeling that separation of the patient from his family is necessary to provide more effective treatment, and the apparent fact that unfamiliarity with partial hospitalization programs lessens the probability of a referral being initiated.

No ready-made solutions to these issues are available at the present time. Inroads into the insurance problem have been made (see Chapter 9), but to date these appear to be isolated achievements. A more generally applicable alternative must be pursued although, at the present time, such a solution does not seem to be imminent.

The problem of increasing the awareness of others regarding the purposes, utility, and cost-effectiveness of partial hospitalization will require long and concentrated effort by proponents of this treatment modality. Goldberg and Perrault (Chapter 9) have suggested some initial steps that can be taken in regard to legislators, insurance carriers, etc. However the investigations cited in this section demonstrate that the educational process must be conducted with equal vigor *within* the mental health profession itself. Recently, McNabola (1975) cited the comparative dearth of information regarding partial hospitalization and urged increased publication in this area. The intervening period has, indeed, witnessed greater activity in this area. However, these efforts must be concentrated in further substantiating the efficacy (both clinical and fiscal) of partial hospitalization treatment and thereby, one hopes, moderating some of the nonclinical biases that have had a detrimental effect on the utilization of this treatment modality.

Conclusions

Since the founding of the first day hospital some 40 years ago, this modality of psychiatric treatment has experienced a slow but steady growth and expansion in several areas including the number of existing programs, the number of patients treated, treatment settings, types of patients treated, and treatment functions. Nevertheless, partial hospitalization has not become the primary form of psychiatric treatment nor has it replaced the traditional inpatient service to any great degree, as some observers expected in its early years of development (Barnes, 1964).

In fact, over the years, partial hospitalization has experienced some difficulty establishing itself as a legitimate component in the continuum of psychiatric care. The roots of this problem appear to be both internal and external. Internally, partial hospitalization, until quite recently, has been able to demonstrate face-validity but only limited empirical support for its contention to be an efficacious treatment form. That evidence, however, is slowly beginning to accumulate and will continue to accrue in the future. *Externally,* partial hospitalization has been subjected to a variety of institutional, organizational, and individual biases that have retarded its development and acceptance; these biases are only now beginning to be recognized and systematically investigated.

These external biases may prove, in the long run, to be the most difficult to alter since they rest on rather pervasive attitudes and traditions. For example, the subjective, nonclinical determinants favoring utilization of inpatient services are supported not only by many clinicians in the psychiatric profession but also by the policies of most third-party funding sources as well. To date, the attitudes and actions of the latter have been more amenable to alterations than the former, although partial hospitalization can by no means claim a victory in the funding area at the present time. The signs are at least encouraging. It remains to be seen, however, whether more general and more realistic fiscal resources can be generated and whether, in turn, the increased ability to generate funds will affect the position of, and attitude toward, partial hospitalization within the psychiatric profession itself.

This chapter has suggested several areas of possible investigation and/or development for partial hospitalization in the future. Undoubtedly, there are other areas of some importance that have not been considered. Also, in the course of this book, some significant questions regarding the treatment modality have been raised. Finally, most of the conceptual and empirical material related to partial hospitalization has been presented in considerable detail.

These data indicate that, in spite of persistent questions that still require investigation, partial hospitalization is an effective treatment modality offering unique advantages over other forms of psychiatric care. Of particular interest are the findings that partial hospitalization can produce results comparable to inpatient treatment at 1-year follow-up at about one-half the cost.

It is, however, still unrealistic to expect partial hospitalization to become the *primary* psychiatric treatment modality. But, it can be hoped and expected that in the near future partial hospitalization will assume a position within the continuum of care that will most effectively utilize the unique advantages of the modality. Certainly, this will be most beneficial to both the providers and the recipients of mental health services.

References

Austin, N. K., Liberman, R. P., King, L. W., & DeRisi, W. J. A comparative evaluation of two day hospitals. *Journal of Nervous and Mental Disease,* 1976, *163,* 253–262.

Barnes, R. Foreword. In R. Epps & L. D. Hanes (Eds.), *Day care of psychiatric patients.* Springfield, Illinois: Charles C Thomas, 1964.

Beigel, A., & Feder, S. L. Patterns of utilization in partial hospitalization. *American Journal of Psychiatry,* 1970, *126,* 101–108.

Bierer, J. Theory and practice of psychiatric day hospitals. *Lancet,* 1959, *2,* 901–902.

Fink, E. B., Heckerman, C. L., & McNeil, D. *Variables in treatment setting determination.* Paper presented at the Annual Conference of the Federation of Partial Hospitalization Study Groups, Boston, 1977(a).

Fink, E. B., Longabaugh, R., & Stout, R. Partial hospital under-utilization. In R. Luber, J. Maxey, & P. Lefkovitz (Eds.), *Proceedings of the annual conference on partial hospitalization.* Boston: Federation of Partial Hospitalization Study Groups, 1977. (b)

Finzen, A. Psychiatry in the general hospital and the day hospital. *Psychiatry Quarterly,* 1974, *48,* 489–495.

Forehand, R., & Atkeson, B. Generality of treatment effects with parents as therapists: A review of assessment and implementation procedures. *Behavior Therapy,* 1977, *8,* 579–593.

Fotrell, E. M. A ten years' review of the functioning of a psychiatric hospital. *British Journal of Psychiatry*, 1973, *123*, 715–717.

Glaser, F. The uses of the day program. In H. Barten & L. Bellak (Eds.), *Progress in community mental health* (Vol. 2) New York: Grune & Stratton, 1972.

Grob, M. C., & Washburn, S. L. A comparative study of factors affecting referral to a psychiatric inpatient or day hospital service. In R. Luber, J. Maxey, & P. Lefkovitz (Eds.), *Proceedings of the annual conference on partial hospitalization.* Boston: Federation of Partial Hospitalization Study Groups, 1977.

Hersen,M., & Bellack, A. S. A multiple baseline analysis of social skills training in chronic schizophrenics. *Journal of Applied Behavior Analysis*, 1976, *9*, 239–245.

Hersen, M., & Luber, R. F. Use of group psychotherapy in a partial hospitalization service: The remediation of basic skill deficits. *International Journal of Group Psychotherapy*, 1977, *27*, 361–376.

Herz, M. I. Day hospitalization as an alternative to inpatient hospitalization. In R. F. Luber, J. Maxey, & P. Lefkovitz (Eds.), *Proceedings of the annual conference on partial hospitalization.* Boston: Federation of Partial Hospitalization Study Groups, 1977.

Hollingshead, A. B., & Redlich, C. F. *Social class and mental illness.* New York: Wiley, 1958.

Krieger, G. Community and service delivery problems in the treatment of de-institutionalized clients. In R. F. Luber, J. Maxey, & P. Lefkovitz (Eds.), *Proceedings of the annual conference on partial hospitalization.* Boston: Federation of Partial Hospitalization Study Groups, 1977.

Liberman, R. P., & Smith, V. A multiple baseline study of systematic desensitization in a patient with multiple phobias. *Behavior Therapy*, 1972, *13*, 597–603.

Lombardo, T. W., & Turner, S. M. Use of thought-stopping to control obsessive ruminations in a chronic schizophrenic patient. *Behavior Modification*, in press.

Luber, R. F., & Hersen, M. A systematic behavioral approach to partial hospitalization: Implications and Applications. *Corrective and Social Psychiatry and Journal of Behavior Technology, Methods, and Therapy*, 1976, *4*, 33–37.

Lubin, B., Hornstra, R., Lewis, R., & Bechtel, B. Correlates of initial treatment assignment in a community mental health center. *Archives of General Psychiatry*, 1973, *24*, 4.

McNabola, M. Partial hospitalization: A national overview. *Journal of the National Association of Private Psychiatric Hospitals*, 1975, *7*, 10–16.

Maxmen, J. S., & Tucker, G. J. The admission process. *Journal of Nervous and Mental Disease*, 1973, *156*, 327–340.

Mendel, W. M., & Rapport, S. Determinants of the decision for psychiatric hospitalization. *Archives of General Psychiatry*, 1967, *20*, 321–328.

Mischler, E. G., & Waxler, N. E. Decision process in psychiatric hospitalization. *American Sociological Review*, 1963, *28*, 576–587.

Parloff, M. B., & Dies, R. T. Group psychotherapy outcome research 1966–1975. *International Journal of Group Psychotherapy*, 1977, *27*, 281–319.

Washburn, S. L., Vannicelli, M., & Scheff, B. J. Irrational determinants of the place of psychiatric treatment. *Hospital & Community Psychiatry*, 1976, *27*, 179–182.

Williams, M., Turner, S., Watts, J., Bellack, A. S., & Hersen, M. Group social skills training for psychiatric patients. *European Journal of Behavioural Analysis and Modification*, 1977, *4*, 223–229.

Wollowitz, J. Utilization review: A proliferation of systems. In R. Luber, J. Maxey, & P. Lefkovitz (Eds.), *Proceedings of the annual conference on partial hospitalization.* Boston: Federation of Partial Hospitalization Study Groups, 1977.

Zwerling, I., & Wilder, J. F., Day treatment for acute psychotic patients. In J. Masserman (Ed.), *Current psychiatric therapies.* New York: Grune & Stratton, 1962.

Author Index

Subject Index